Paul Şerban Agachi, Zoltán K. Nagy,
Mircea Vasile Cristea, and Árpád Imre-Lucaci

Model Based Control

Related Titles

Robert Haber, Ruth Bars, Ulrich Schmitz

Predictive Control in Process Engineering

2007
ISBN-10: 3-527-31492-X
ISBN-13: 978-3-527-31492-8

Frerich J. Keil (ed.)

Modeling of Process Intensification

2007
ISBN-10: 3-527-31143-2
ISBN-13: 978-3-527-31143-9

Alexandre C. Dimian, C. Sorin Bildea

Chemical Process Design
Computer-Aided Case Studies

2007
ISBN-10: 3-527-31403-2
ISBN-13: 978-3-527-31403-4

Luis Puigjaner, Georges Heyen (eds.)

Computer Aided Process and Product Engineering
2 Volumes

2006
ISBN-10: 3-527-30804-0
ISBN-13: 9783-527-30804-0

Sebastian Engell (ed.)

Logistic Optimization of Chemical Production Processes

2006
ISBN-10: 3-527-30830-X
ISBN-13: 978-3-527-30830-9

Kai Sundmacher, Achim Kienle, Andreas Seidel-Morgenstern (eds.)

Integrated Chemical Processes
Synthesis, Operation, Analysis, and Control

2005
ISBN-10: 3-527-30831-8
ISBN-13: 978-3-527-30831-6

Zivorad R. Lazic

Design of Experiments in Chemical Engineering
A Practical Guide

2004
ISBN-10: 3-527-31142-4
ISBN-13: 978-3-527-31142-2

Paul Şerban Agachi, Zoltán K. Nagy,
Mircea Vasile Cristea, and Árpád Imre-Lucaci

Model Based Control

Case Studies in Process Engineering

WILEY-VCH
VCH

WILEY-VCH Verlag GmbH & Co. KGaA

The Authors

Prof. Dr. Paul Şerban Agachi
Babeş-Bolyai University
Dept. of Chemical Engineering
M. Kogalniceanu 1
400084 Cluj-Napoca
Romania

Prof. Dr. Zoltán K. Nagy
Loughborough University
Chemical Engineering Department
Loughborough, LE11 3TU
United Kingdom

Prof. Dr. Mircea Vasile Cristea
Babeş-Bolyai University
Dept. of Chemical Engineering
M. Kogalniceanu
400084 Cluj-Napoca
Romania

Prof. Dr. Árpád Imre-Lucaci
Babeş-Bolyai University
Dept. of Chemical Engineering
M. Kogalniceanu
400084 Cluj-Napoca
Romania

Cover Adam Design, Weinheim

■ All books published by Wiley-VCH are carefully produced. Nevertheless, authors, editors, and publisher do not warrant the information contained in these books, including this book, to be free of errors. Readers are advised to keep in mind that statements, data, illustrations, procedural details or other items may inadvertently be inaccurate.

Library of Congress Card No.:
applied for

British Library Cataloguing-in-Publication Data
A catalogue record for this book is available from the British Library.

Bibliographic information published by the Deutsche Nationalbibliothek
The Deutsche Nationalbibliothek lists this publication in the Deutsche Nationalbibliografie; detailed biliographic data are available in the Internet at <http://dnb.d-nb.de>.

Typesetting   Dörr + Schiller GmbH, Stuttgart
Printing   betz-druck GmbH, Darmstadt
Binding   Litges & Dopf, Heppenheim

Printed in Germany
Printed on acid-free paper

ISBN-13:   978-3-527-31545-1
ISBN-10:   3-527-31545-4

Table of Contents

Agachi/Nagy/Cristea/Imre-Lucaci. *Model Based Control*
Copyright © 2006 WILEY-VCH Verlag GmbH & Co. KGaA, Weinheim
ISBN: 3-527-31545-4

Preface

Today, the process industries are challenged by a dynamic and difficult-to-predict market, leading to consumer-oriented manufacturing and tailor-made products. Those times which were characterized by constant mass production are over and, due largely to globalization, manufacturers are now obliged to react to exquisite demands in a very short time, with a wide variety of quality specifications, and with competitive prices. Moreover, environmental regulations have placed stricter limitations on the production of pollutants. Some actors in the field of process industries complain that all their benefits are consumed in satisfying the "fiercely enforced" environmental regulations. Indeed, environmental considerations have led to the design of smaller storage capacities in order to reduce the risk of major accidents. Likewise, large increases in energy costs (today's costs for oil and natural gas prices have doubled since 2004) have encouraged the design of thermally highly integrated plants for which dynamic stability and control offer real challenges; additionally, every last morsel of saved energy or raw materials demands a sustainable development approach, and this is reflected not only in costs and prices but also in the preservation of reserves.

The above-mentioned facts explain the current interest in new process control techniques that enable the operation of flexible, high-performance, optimized and variable-capacity production lines. The improved control of chemical processes means greater production for the same production equipment, improved product quality, reduction of waste and pollution, and reduced energy consumption. Advanced control is one of the most important ways in which the production situation can be improved, and model-based control offers a very direct and feasible solution for an appropriate operation.

The roots of my interest in Model Predictive Control (MPC) stem from my days at Caltech, Chemical Engineering Department where, during 1991 and 1992, I worked in the group of Professor Manfred Morari. This American experience was extraordinary, because it was at the start of a new era of the scientifically oriented software packages Matlab and Simulab (currently Simulink), and also at the start of MPC implementation on a larger scale and on complicated industrial processes. At the time, nobody at Caltech had any great knowledge of using this software correctly, and many nights and weekends were spent attempting to find rapid

Agachi/Nagy/Cristea/Imre-Lucaci. *Model Based Control*
Copyright © 2006 WILEY-VCH Verlag GmbH & Co. KGaA, Weinheim
ISBN: 3-527-31545-4

solutions for complicated models of the Fluid Catalytic Cracking Unit (FCCU), using a Matlab package that ran slowly, without a compiler at that time, together with my colleagues Argimirio Secchi (now Professor of Chemical Engineering in Brazil) and Richard Braatz (now Professor of Chemical Engineering in the US). I owe them a lot, and I thank them in this way. Together with our Italian colleague Bruno Donno, and students John Bomberger and Iftiqar Huq, we have succeeded in elaborating the first MPC controller in Simulink, and have also identified some approximate rules for the optimal tuning of MPC parameters. I remember the weekly discussions with Professor Morari which contributed greatly to the progress of knowledge in this field, and I thank him sincerely for the opportunity given not only at that time but also subsequently as Professor at ETH, Zürich. My collaboration with Chevron Research and Development during this project was remarkable when, together with Ron Soerensen, we tried to implement the FCCU mathematical model and the MPC theory in their refinery at El Segundo. Later, collaboration with Professor Frank Allgöwer, former colleague and friend at Caltech, and now Head of the Institute for Systems Dynamics at the University of Stuttgart, was also highly profitable from an MPC research point of view.

On returning to Romania, my students and now colleagues – some of whom are coauthors of this book – took up the challenge and continued with great enthusiasm in this challenging field of MPC. Together, we have applied the elaborated theory (which we have enriched with modesty and patience) to several processes, beginning with our laboratory of Automatic Control in Process Engineering. We have largely used our data bank of Chemical Process Dynamic Models, the content of which covers an activity period of more than twenty years.

The aim of this book is first to present very briefly the general issues related to MPC, which have long been reported in the literature, in addition to the innovations of our group in specific algorithms. Second – and we consider this is the real "added value" of the book – we have presented several complicated applications of model-based control to several processes, ranging from the petrochemical industry to the ceramic or chlorine industries and to bioengineering. We have added – proudly I may say – the "real" laboratory application of MPC of a pilot distillation column, which in reality has demonstrated the superiority of model-based techniques in relation to classic control methods.

The book is structured in three parts:
- Chapter 1 – the Introduction – treats issues related to the concepts of process control, the progress of modern control theory, and its applications. The evolution was strongly related to – and even determined by – the progress of the computer industry. The features of processes controlled by advanced control techniques are briefly presented. Finally, a classification of Advanced Control Techniques is introduced.
- In Chapter 2, MPC is directed towards Internal Model Control (IMC), Linear MPC (LMPC) and Nonlinear MPC (NMPC), with the history and basics theory of these types of algorithm. The final section of this chapter provides the methods for tuning MP controller parameters.

- Chapter 3, which is the most elaborate, includes the Case Studies. This chapter contains several applications of MPC, each starting with the elaboration of a dynamic model shaped in a form to be appropriate for control. The chapter includes original dynamic models of the FCCU, polyvinyl chloride reactor, yeast fermentation reactor, distillation column, ceramic dryer, amalgam, and ion-exchange membrane (IEM) electrolyzers which are not normally (or are perhaps rarely) found in the literature. These models have been elaborated in the Department of Chemical Engineering and Oxide Material Science of the University "Babeş-Bolyai" in Cluj-Napoca, Romania, more precisely in the group of Computer-Aided Process Engineering. They have been developed on the basis of both literature and industrial data, using the experience of the group members, and are aimed at improving process operation by the implementation of model-based control techniques. The chapter also provides the MPC solutions that are compared with other control solutions.

We would like to emphasize that process control continues to be a research battlefield where challenging industrial problems await their particular solutions. Nonlinear control strategies, artificial intelligence techniques, process integration and process optimization are important research areas with an increasing number of industrial products and commercial applications.

This book is addressed to a broad category of technical staff, ranging from researchers, plant engineers and technicians to students and academic personnel with control engineering or chemical engineering backgrounds, and who are interested in the exciting intertwined fields of process engineering, industrial economics, process modeling, optimization, artificial intelligence and process control.

We are very much indebted to the Wiley-VCH Editing House, which showed interest in the topics treated in this book, and particularly to Ms. Karin Sora and Ms. Waltraud Wüst who provided valuable advice during our writing of the book, and also during its editing and presentation to the market.

Paul Şerban Agachi
Professor of Process Control
Department of Chemical Engineering and Oxide Material Science
University "Babeş-Bolyai", Cluj-Napoca, Romania
Cluj-Napoca, July, 2006

1
Introduction

Process control, as it has been known for many years, was first developed in the process industries. Although starting with local measurement devices, the system has since progressed to centralized measurement and control (central control rooms) and computer hierarchical/plantwide control. Recent developments in process control have been influenced by improvements in the performance of digital computers suitable for on-line control. Moreover, while the performance of these units has improved significantly, their prices have fallen drastically. The price trends for small but more sophisticated minicomputers, despite the inclusion of more reliable electronics and increasing inflation, is shown graphically in Figure 1.1. By having high speeds of operation and storage capacities, the process computer can be used effectively in process control due to its insignificant capital cost. Once in place, the computer is usually operated in a timesharing mode with large numbers of input/output operations, so that the central processing unit (CPU) is typically in use only for about 5 % of the time. Thus, many industrial plants have 95 % of the computing power of a highly capable minicomputer programmable in a high-level language such as Fortran, C, Visual Basic, LabVIEW, etc., and available for implementing sophisticated computer-controlled schemes.

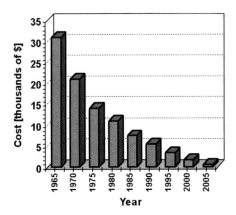

Figure 1.1 Price trends for real-time minicomputers.

Agachi/Nagy/Cristca/Imre-Lucaci. *Model Based Control*
Copyright © 2006 WILEY-VCH Verlag GmbH & Co. KGaA, Weinheim
ISBN: 3-527-31545-4

During the same time, *modern control theory* has undergone intense development, with many successful applications covering many areas of the industry. Most recently, several process control research groups have applied new sophisticated control algorithms and schemes to simulated, laboratory-scale – and even full-scale – processes. As a consequence, it is necessary for the process control engineer to design an economically optimal process control scheme based on a judicious comparison of the available control algorithms. The aim of this book is to offer assistance in this respect, and to provide a brief introduction to the theory and practice of the most important modern process control strategies. This is achieved by using industrially relevant chemical processes as the subjects of control performance studies.

1.1
Introductory Concepts of Process Control

A *control system* is a combination of elements which act together in order to bring a measured and controlled variable to a certain, specific, desired value or trajectory termed the "set/point of reference". The basis for an analysis of such a system is the foundation provided by linear system theory, which assumes a linear cause–effect relationship for the components of a system. Therefore, a component or *process* to be controlled can be represented by a block, as shown in Figure 1.2. The output variables are the "interesting" ones (technological parameters, yield, etc.), while the input variables are those which influence the outputs (e.g., mass or energy flows, environmental variables, etc.). Figure 1.2 illustrates the different types of *input* and *output* parameters used in the development and study of control algorithms. We refer to a variable as an *input* if its value is determined by the "environment" of the system to be controlled. We distinguish *disturbance* inputs and *manipulated* or *control* inputs. We are free to adjust the later but not the former. Variables, the values of which are determined by the state of the system, are referred to as *outputs* – some of these are measured, but others are not. *Controlled* variables must be maintained at specified setpoints. *Associated* variables are only required to stay within certain bounds, their exact value within bounds being of little interest.

An *open-loop* control system uses a controller or control actuator in order to obtain the desired response, as shown in Figure 1.3. In contrast to an open-loop control system, a closed-loop control system uses an additional measure of the

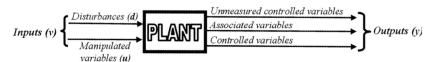

Figure 1.2 Definition of input and output variables considered for control system design.

Figure 1.3 Open-loop control system.

actual output in order to compare the actual output with the desired output response. The measure of the output is called the *feedback signal*. A simple *closed-loop feedback control system* is shown in Figure 1.4. A standard definition of a feedback control system is as follows: *A feedback control system is a control system that works to maintain a prescribed relationship between one system variable and another by comparing functions of these variables and using the difference as a means of control.*

A feedback control system often uses a function of a prescribed relationship between the output and the reference input to control the process. Often, the difference between the output of the process under control and the reference input is amplified and used to control the process, so that the difference is continually reduced. The feedback concept has been the foundation for control system analysis and design.

Classical control theory is essentially limited to single-input single-output (SISO) systems described by linear differential equations with constant coefficients (or their corresponding Laplace transforms). However, the so-called *modern* control theory has developed to the point where results are available for a wide range of general multivariable systems, including those described by linear, variable-coefficient differential equations, nonlinear differential equations, partial differential, and integral equations.

The results of modern control theory include the so-called *optimal control theory*, which allows the design of control schemes, which are optimal in the sense that the controller performance minimizes some specified cost functional.

In addition to controller design, modern control theory includes methods for process identification and state estimation. *Process identification* algorithms have been developed for determining the model structure and estimating the model parameters, either off-line or adaptively on-line. These are useful both in the initial control system design and in the design of *adaptive control systems* which respond to such changes in the process characteristics. These might arise, for example, with the fouling of heating exchanger surfaces or the deactivation of catalyst in chemical reactors. *State estimation* techniques are on-line methods either for estimating

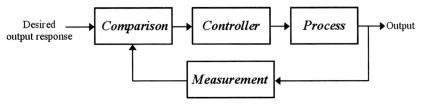

Figure 1.4 Closed-loop feedback control system.

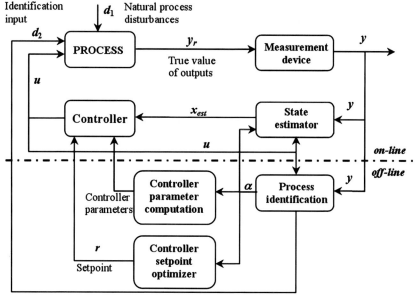

Figure 1.5 A comprehensive advanced computer control scheme.

system state variables which are not measured, or for improving the quality of all the state-variable estimates in the presence of measurement errors. In those processes where some sensors are not available, or are too expensive to be installed, on-line state estimation can be of significant practical importance.

The way in which all the components of a comprehensive computer process control scheme might fit together for a particular process is illustrated in Figure 1.5. Such a control scheme would consist of the following parts:

- The *Process*, which responds to control inputs u; to natural process disturbances d_1; and to special input disturbances d_2 used for identification. The true process state x is produced, but this is seldom measured either completely or precisely.
- *Measurement devices*, which usually are able to measure only a few of the states or some combination of states, and are always affected by measurement errors. The measurement device outputs y are fed to the –
- *State estimator*, which uses the noisy measurements y along with a process model to reconstruct the best possible process state estimates x_{est}. The process state estimates are passed to the –
- *Controller*, which calculates what control actions must be taken based on the state estimates x_{est}, the setpoints r (which themselves may be the subject of process optimization), and the controller tuning parameters. The controller parameters can be calculated either off-line or adaptively on-line, based on current estimates of the model parameters. The process model parameters must be determined from the –

- *Process identification block*, which takes user measurements from the process as raw data y (and may choose to introduce experimentally designed input disturbances d_1) in order to identify the process model parameters α. If the parameters are time-invariant, the identification is unique; however, if the process changes with time, then the identification scheme must be activated periodically to provide adaptation to changing conditions.

In most applications only a few of the components of this control structure are required.

1.2
Advanced Process Control Techniques

1.2.1
Key Problems in Advanced Control of Chemical Processes

The main features of chemical processes that cause many challenging control problems [1–3] are shown schematically in Figure 1.6.

1.2.1.1 Nonlinear Dynamic Behavior
Nonlinear dynamic behavior of chemical processes causes one of the most difficult problems in designing control systems. In the case of linear, lumped parameter systems, a very general model in the time domain form can be written as:

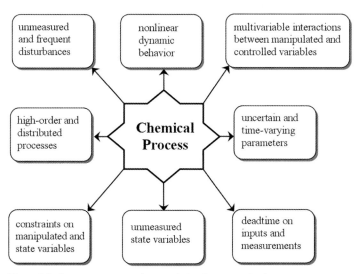

Figure 1.6 Common process characteristics, important in the choice of control strategy.

$$\frac{dx}{dt} = A \cdot x + B \cdot u + \Gamma \cdot d, \; x(t_0) = x_0 \tag{1.1}$$

$$y = C \cdot x \tag{1.2}$$

where x, u, y, and d are the vectors of states, controls (manipulated variables), outputs and disturbances, respectively. The state space matrices, A, B, C, and Γ, can be either constant or time-varying. For systems in Laplace transform domain, involving transfer functions the model can be represented in the form:

$$\bar{y}(s) = G(s) \cdot \bar{u}(s) + G_d(s) \cdot \bar{d}(s) \tag{1.3}$$

with:

$$G(s) = C(sI - A)^{-1}B \tag{1.4}$$

$$G_d(s) = C(sI - A)^{-1}\Gamma \tag{1.5}$$

These representations show the linear dependence between the manipulated inputs (u) and outputs (y); that is, a certain Δu variation will cause, in a certain time t, a linear proportional variation of y ($\Delta y = \alpha_t \Delta u$). Assuming that $d = 0$, (there is no unmeasured disturbance), once the linear model was identified (the coefficients of A, B, C, and Γ were determined), the trajectory of the outputs y can be predicted for any changes of the manipulated inputs, Δu, at any time using the linear model in one of the forms described above [Eqs. (1.1)–(1.2) or (1.3)–(1.5)].

In the case of nonlinear processes, there is no linear dependence between control variables and states (or manipulated variables), so that Eqs. (1.1)–(1.5) are no longer valid, and for predictions a much more general model must be used. Mathematically, a general nonlinear process model can be represented as follows:
- dynamic modeling equations:

$$\frac{dx}{dt} = f(x, u, q, d) \tag{1.6}$$

with the *initial conditions*:

$$x(t_0) = x_0 \tag{1.7}$$

- algebraic equations (equilibrium relationship, etc.)

$$0 = g_1(x, u, q) \tag{1.8}$$

- state-output relationship:

$$y = g_2(x) \tag{1.9}$$

where x are state variables, u are manipulated variables, q are parameters, d are measured and unmeasured disturbances, and y is the output (measured) variable.

In contrast to the case of a linear model, for nonlinear process model there is generally no analytical solution and the prediction must be made by numerically solving the model. Consequently, for nonlinear process models, computational demand is much higher than for the linear ones.

1.2.1.2 Multivariable Interactions between Manipulated and Controlled Variables

It is commonly believed that for SISO systems, well-tuned proportional, integral, derivative (PID) controllers work as well as model-based controllers, and that PID controllers are more robust to model errors. The offset-free constrained linear quadratic (LQ) controller for SISO systems, may be implemented in an efficient way so that the total controller execution time is similar to that of a PID [4].

Unfortunately, most multivariable systems have significant coupling between outputs and controls, and these pose great difficulties in control system design. In the case of multiple input-multiple output (MIMO) systems, any manipulated input can have effects on more outputs. Thus, the choice of appropriate control loops with the best control performances demands detailed study and can be very difficult. One of the main advantages of most advanced control strategies is that they can explicitly handle the multivariable interactions. Due to their multivariable nature, advanced control strategies – such as model predictive control techniques (MPC) – allow the control problem to be addressed globally. Thus, one must determine only the best set of manipulated inputs for a certain set of controlled outputs, and there is no need for detailed study of the individual interactions between the inputs and outputs. However, in the choice of the best control set, a study of the interaction problem for a certain MIMO system application is always useful.

1.2.1.3 Uncertain and Time-Varying Parameters

Most chemical processes are characterized by having uncertain and/or time-varying parameters. Time-varying parameters are common for batch and semibatch processes, when it is clear that most of the thermodynamic and physico-chemical properties of the system vary with time. Moreover, even for continuous processes when deviations from steady-state are frequent and the process variables vary in a wide operating range, the dependence of parameters on time should be taken into consideration.

For linear time-varying processes the state-space representation of the model [Eqs. (1.1)–(1.2)] is still valid, but in this case the elements of the state-space matrices A, B, C, and Γ are functions of time. The general model of nonlinear systems expressed by Eqs. (1.6)–(1.9), by its mathematical form takes explicitly into consideration the variation in time of the process parameters.

Among the multitude of parameters of a chemical process model, a significant number cannot normally be determined accurately, and this will lead to model/plant mismatch. The importance of uncertainties is increasingly being recognized by control theoreticians; consequently, they are being included explicitly in the formulation of control algorithms. MPC can handle model/plant mismatch in its closed loop, feedback form, by continuously adjusting the uncertain parameters so that the difference between the current measurements and the prediction from the previous step is a minimum.

1.2.1.4 Deadtime on Inputs and Measurements

One especially important class of systems in chemical process control is that of having time delays. This class of dynamic systems arises in a wide range of applications, including paper making, chemical reactors, or distillation. The principal difficulty with time delays in the control loop is the increased phase lag, which leads to unstable control system behavior at relatively low controller gains. This limits the amount of control action possible in the presence of time delays. In multivariable time-delay systems with multiple delays, these problems are even more complex. In these problems, the normal control difficulties due to loop interactions are complicated by the additional effect of time delays. Consequently, these aspects must be taken into consideration. The properly designed process model in its general nonlinear form expressed by Eqs. (1.6)–(1.9) explicitly involves time delays; however, to emphasize this feature in the literature one can often find the following general mathematical expression of the nonlinear model:

$$\frac{dx}{dt} = f(x(t - \Phi_x), u(t - \Theta), q, d) \tag{1.10}$$

$$x(t_0) = x_0 \tag{1.11}$$

$$0 = g_1(x, u, q) \tag{1.12}$$

$$y = g_2(x(t - \Phi_y)) \tag{1.13}$$

where Θ is the deadtime between manipulated and state variables, Φ_y is the deadtime between manipulated and output variables, and Φ_x is the deadtime on state variables.

From this model it can be seen that, in a very general form, deadtime can be included on process inputs and control variables as well as on unmeasured states.

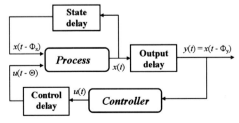

Figure 1.7 The structure of a controlled system having state, control and output delays.

For example, time-delay can be due to transport (flow through pipes) or measurement delays (analytical instrumentation, etc.). A good example of a multivariable time-delay nonlinear system with multiple delays is that of the distillation process.

A very general representation of systems having delays in the control variables $u(t)$, state variables $x(t)$ and output variables $y(t)$ is illustrated in Figure 1.7.

1.2.1.5 Constraints on Manipulated and State Variables

In chemical process control, constraints on state variables usually arise due to technological specifications, while those on manipulated variables are caused generally by the control hardware restrictions as well as the control system characteristics. For example, in systems with time delay in the control loop, the controller gain must be limited in order to avoid unstable behavior. In practice, the operating point of a plant that satisfies the overall economic goals of the process usually lies at the intersection of constraints. Therefore, in order to be successful, any control system must anticipate constraint violations and correct them in a systematic manner. Violations of the constraints must not be allowed while the operation is kept close to these constraints.

Constraints on manipulated and output/state variables can be expressed mathematically as follows:

$$y_{min} \leq y \leq y_{max} \tag{1.14}$$

$$u_{min} \leq u \leq u_{max} \tag{1.15}$$

$$|\Delta u| = \Delta u_{max} \tag{1.16}$$

where the limits of the state/output variables (y_{min}, y_{max}) and those of control inputs/manipulated variables (u_{min}, u_{max}, Δu_{max}) can be either constant or time-varying.

The usual practice in process control is to ignore the constraint issue at the design stage and then to "handle" it in an "ad hoc" way during the implementation. Therefore, these control structures are very system-specific, and their cost cannot be spread over a large number of applications, implying high design cost. Advanced control techniques usually provide intelligent methodologies to handle constraints in a systematic manner during the design and implementation of the control.

1.2.1.6 High-Order and Distributed Processes

On many occasions, the modeling of chemical processes leads to very high-order models. Although the general nonlinear model expressed by Eqs. (1.6)–(1.9) is a specific formulation for simple, low-order, lumped parameter systems which can be described by ordinary differential equations, this form can be also used for high-order and/or distributed parameter systems. Generally, in the case of high-order systems, an nth order differential equation can be described by a system of n first-order differential equations by introducing $n-1$ fictitious state variables. For example, for a high-order system with one state variable, described by the nth order differential equation below:

$$a_n \frac{d^n x}{dt^n} + a_{n-1} \frac{d^{n-1} x}{dt^{n-1}} + \ldots + a_1 \frac{dx}{dt} + a_0 x + b = 0 \tag{1.17}$$

is equivalent with the following system of n first-order differential equations:

$$\frac{dx}{dt} = x_1$$

$$\frac{dx_1}{dt} = x_2 \tag{1.18}$$

$$\vdots$$

$$\frac{dx_{n-1}}{dt} = -\frac{a_{n-1}}{a_n} x_{n-1} - \ldots - \frac{a_1}{a_n} x_1 - \frac{a_0}{a_n} x - \frac{b}{a_n}$$

where $x_1, \ldots, x_n-1$ are fictitious states.

In this way, the general model described by Eqs. (1.6)–(1.9) can be extrapolated for high-order systems. However, for a MIMO system every high-order equation must be decomposed in a system of ordinary differential equations, and this – usually in the case of very high-order MIMO systems – can cause computational difficulties. For this reason, a model reduction is recommended in the case of high-order systems.

Distributed parameter systems are distinguished by the fact that the states, controls and outputs may depend on spatial position. Thus, the natural form of the system model is represented by partial differential equations or integral equations.

1.2.1.7 Unmeasured State Variables and Unmeasured and Frequent Disturbances

In most industrial processes, the total state vector can seldom be measured, and the number of outputs is much smaller than the number of states. In addition, the process measurements are often corrupted by significant experimental error, and the process itself is subject to random, unmodeled upsets. Both, unmeasured state variables and unmeasured disturbances can lead to a substantial model/plant mismatch, which appears as a reduction in quality control. However, each of these difficulties individually causes a very challenging control problem (according to most control specialists, the most important problem in MPC design): the consequences for both problems are differences between the predicted (y_p) and measured (y_m) outputs. Thus, the effects of the unmeasured disturbances can be included in the model error caused by the unmeasured state variables, and treated in the model/plant mismatch problem as a global, additive disturbance. Because unmeasured state variables and unmeasured disturbances manifest themselves in the quality of the predictions, which actually underlines the MPC strategies, the state estimation is an essential problem in practical NMPC applications.

1.2.2
Classification of the Advanced Process Control Techniques

The chemical process industry is characterized as having highly dynamic and unpredictable marketplace conditions. For example, during the course of the past 15 years we have witnessed an enormous variation in crude and product prices. The demands for chemical products also vary widely, imposing different production yields. It is generally accepted that the most effective means of generating the highest profit from plants, while responding to marketplace variations with minimal capital investment, is provided by integrating all aspects of automation of the decision-making process [6], which are:
- *Measurement.* The gathering and monitoring of process measurements via instrumentation.
- *Control.* The manipulation of process degrees of freedom for the satisfaction of operating criteria. This typically involves two layers of implementation: the single loop control which is performed via analogue controllers or rapid sampling digital controllers; and the overall control performed using real-time computers with relatively large CPU capabilities.
- *Optimization.* The manipulation of process degrees of freedom for the satisfaction of plant economic objectives. This is usually implemented at a rate such that the controlled plant is assumed to be at steady state. Therefore, the distinction between control and optimization is primarily a difference in implementation frequencies.
- *Logistics.* The allocation of raw materials and scheduling of operating plants for the maximization of profits and the achievement of the company's program.

Each of these automation layers plays a unique and complementary role in allowing a company to react rapidly to changes. Therefore, one layer cannot be effective without the others. In addition, the effectiveness of the whole approach is only possible when all manufacturing plants are integrated into the system.

Although, in the past, the maintenance of a stable operation for the process was the sole objective of control systems, this integration imposes more demanding requirements. In the process industries, control systems must satisfy one or more of the following practical performance criteria:
- *Economic.* These can be associated with either maintaining process variables at the targets dictated by the optimization phase, or dynamically minimizing an operating cost function.
- *Safety and environmental.* Some process variables must not violate specified bounds for reasons of personnel or equipment safety, or because of environmental regulations.
- *Equipment.* The control system must not drive the process outside the physical limitations of the equipment.
- *Product quality.* Consumer specifications on products must be satisfied.
- *Human preference.* There exist excessive levels of variable oscillations or jaggedness that the operator will not tolerate. There can also be preferred modes of operation.

In addition, the implementation of such integrated systems is forcing the processes to operate over an ever-wider range of conditions. As a result, we can state the control problem that any control system must solve as follows [5]:

> "On-line update the manipulated variables to satisfy multiple,
> changing performance criteria in the face of changing plant
> characteristics."

Today, the entire spectrum of process control methodologies in use is faced with the solution of this problem. The difference between these methodologies lies in the particular assumptions and compromises made in the mathematical formulation of performance criteria, and in the selection of a process representation. These are made primarily to simplify the mathematical problem so that its solution fits the existing hardware capabilities. The natural mathematical representation of many of these criteria is in the form of dynamic objective functions to be minimized and of dynamic inequality constraints. The usual mathematical representation for the process is a dynamic model with its associated uncertainties.

At present, there is an important number of advanced control techniques using either specific algorithms for particular systems, or very general methods with a wide application area and well-developed theory. A classification of these techniques is difficult because many of the algorithms are very similar, being obtained

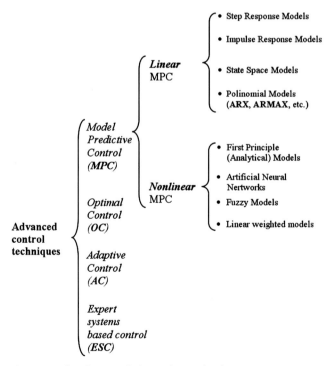

Figure 1.8 Classification of advanced control techniques.

from some more general methods with usually minor changes with regard to, for example, the performance criteria, optimization method, prediction horizon, and constraint handling. However, all of these algorithms have a common feature: all are based on a process model, described in different ways. The proposed classification, based on this feature, is presented in Figure 1.8. According to this, the advanced control techniques can be classified first in four conceptually different categories. The first and most important approach, the Model Predictive Control (MPC), can be classified further, for example, according to different model types used for prediction in the controller. This feature is usually the most significant difference among MPC algorithms.

References

1 Bequette, B. W., Nonlinear Control of Chemical Processes – A Review, *Ind. Eng. Chem. Res.*, **1991**, *30*, 1391–1413.

2 Clarke, D. W., Mohtadi, C., Tuffs, P. S., Generalized predictive control – Part I. The basic algorithm, *Automatica*, **1987**, *23*(2), 137–148.

3 Kothare, M. V., Balakrishnan, V., Morari, M., Robust constrained model predictive control using linear matrix in-equalities, *Automatica*, **1996**, *32*(10), 1361–1379.

4 Pannocchia, G., Laachi, N., Rawlings, J. B., A candidate to replace PID control: SISO-constrained LQ control, *AIChE Journal*, **2005**, *51*(4), 1178–1189.

5 Garcia, C. E., Prett, D. M., Morari, M., Model Predictive Control: Theory and practice – A survey, *Automatica*, **1989**, *25*, 335–347.

2
Model Predictive Control

2.1
Internal Model Control

It is an obvious fact that control system design is fundamentally determined by the steady state and dynamic behavior of the process to be controlled. It is interesting to investigate the way in which the process characteristics influence the controller structure and tuning parameters. Although the transfer of process behavior into the controller structure is an intrinsic part of every control algorithm design developed along the historical evolution of the control systems, this procedure was not always consciously performed. In the early age of automatics it led to ad hoc control design methodologies able to cover only particular cases and lacking comprehensiveness of the control problem approach. The situation improved when the ability to build valid mathematical models was developed in association with enhanced simulation tools. The Internal Model Control (IMC) viewpoint appeared as an alternative to the traditional feedback control algorithm, marking a qualitative step for direct linking of the process model with the controller structure. The name itself denotes the fact that the process model becomes an explicit part of the controller. Although the model-based control techniques have been largely extended and gained prominence during the past decades, major steps are expected in the future, especially for the nonlinear case.

The dawn of the structure that became later the IMC philosophy may be found during the late 1950 s when the investigations of Newton, Gould and Kaiser [1] pointed out the transformation of the closed-loop structure into an open one in order to develop a H_2-optimal controller, and when Smith [2] proposed a predictor to "eliminate" the dead time from the control loop. Frank [3] was the first to anticipate the value of the control structure having in parallel the model and the plant. Brosilow [4], with his inferential control system, also addressed the IMC structure. However, it was Morari and Garcia [5–7] who brought the major contribution for the advance of the new control structure and for revealing it in distinct theoretical framework. The studies of Morari greatly enlarged the IMC design methods unifying the concepts referring to this control structure, and rendering the name that is today recognized by the control community.

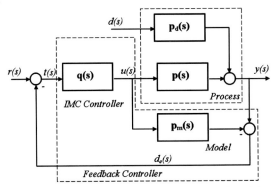

Figure 2.1 Schematic representation of the Internal Model Control (IMC) structure.

The IMC approach to controlling a process has, at its basics, a very human style. When the operator, in manual mode, attempts to maintain a controlled variable close to a desired setpoint, he or she performs a simple calculation based on their intuitive representation (model) of the process in order to set the proper value of the manipulated variable. The operator calculates the difference between the actual value of the controlled output and an estimation (prediction) of the effect of the intended value of the manipulated variable on the plant output. The calculation of this difference is the basic information on which relies the decision to set the amplitude of the manipulated variable change that is sent to the plant. In fact, the operator determines the necessary change of the manipulated variable on a model-based estimation (performed in their mind) of the disturbance. Successive iterations of this procedure lead to a desired behavior of the controlled variable. The same fundamental control approach serves as the core of the internal model control.

A schematic representation of the IMC structure is presented in Figure 2.1 [8,9].

In Figure 2.1, $p(s)$ represents the process (plant) transfer function on the manipulated variable to the controlled variable path, $p_d(s)$ the process transfer function on the disturbance to the controlled variable path, $p_m(s)$ the mathematical model (as transfer function) of the process, and $q(s)$ the transfer function of the IMC controller.

As may be observed from the block diagram of the IMC structure, there are two parallel paths starting from the manipulated variable $u(s)$: one passing through the real process $p(s)$; and one passing through the model of process $p_m(s)$. The role of the parallel containing the model $p_m(s)$ is to make possible the generation of the difference between the actual process output $y(t)$ and an estimation (model-based prediction) of the manipulated variable effect on the process output. Assuming that the process model is a perfect representation of the real process $p_m(s) = p(s)$, the difference $d_e(s)$ represents the estimated effect of the disturbances (both measured and unmeasured) on the controlled variable. If the process model is not perfect, the difference $d_e(s)$ includes both the effect of disturbances on the output variable and the process–model mismatch. The feedback of the control system is zero when the model is perfect and there are no disturbances, resulting in a control loop being open-loop. This fact leads to one of the most important conceptual usefulness of the IMC structure referring to the stability issue. Namely, that the IMC control loop is stable if – and only if – the process $p(s)$ and the IMC controller $q(s)$ are stable, provided that the model is a perfect representation of the process $p_m(s) = p(s)$ and the process is stable. It is only necessary to focus on IMC controller design for avoiding difficulties associated with usual feedback stability problems [8,9].

Considering again the control structure of Figure 2.1, the disturbance estimation $d_e(s)$ may be regarded as a correction for the setpoint $r(s)$ in order to generate an improved target variable $t(s)$ that allows the IMC controller to produce the manipulated variable able to eliminate the disturbance estimation. It is also interesting to note that the IMC controller acts as a feed-forward controller having the important incentive of counteracting the effect of the unmeasured disturbances, as the feedback signal also represents the estimation of their effect on the process output and the controller setpoint is adjusted consequently.

Even when the model of the process is not perfect and the model error determines a feedback signal in the true sense, it is possible to find (detune) the ideal IMC controller $q(s)$ to assure stability, with the only condition that the process is stable by itself.

2.2
Linear Model Predictive Control

The Model Predictive Control (MPC) concept has its roots located about four decades ago when Zadeh and Whalen [10] realized the connection between minimum time optimal control and linear programming, and when Propoi [11] proposed for the first time the moving horizon approach (which is one of MPC's main features). This was first denoted as "Open Loop Optimal Control", but subsequently ignored until the late 1970 s. It was Richalet who restored the neglected MPC importance, reporting successful applications of the so-named "heuristic model predictive control" [12]. Shortly afterwards, Cutler and Ramaker [13] on one side and Prett, Gillette [14] on the other side, developed the incoming MPC concept, naming it "Dynamic Matrix Control", and applying it to oil cracking process control. Dynamic Matrix Control soon became one of the most tempting control strategies [15–17], and several applications soon appeared, including MPC of the steam boiler, pulp and paper processing, glass oven, or continuous and batch reactors. The interest in MPC has grown continuously with in time [18–21], such that today a large number of companies offer hardware and software products for its application. Indeed, the success of application has made certain a rapid payback of the implementation costs.

The large majority of successful MPC applications address the case of multivariable control in the presence of constraints, motivating its extensive distribution for applications where traditional control usually comes close to its limits [22].

Adaptive or robust MPCs have attracted much interest both in academia and industry [23–25]. A rigorous and general theory, as well as an associated efficient computational methodology, was developed that addresses the question of when and what linear control is adequate for a nonlinear process [26].

As its name suggests, MPC relies on an explicit representation of the process to be controlled, bringing the model of the process "inside" the control algorithm in a straightforward manner. The core of MPC does not change with the particular form of the model description; rather, it results in different mathematical formulations and computation load [27,28]. Therefore, linear MPC (LMPC) may be found both in input-output [29,30] or state space [31,32] formulations, for SISO or MIMO cases, stable or integrating behavior [33–35], and explicit or implicit formulation of the control law [36,37].

In order to introduce the main aspects of linear MPC a simple discrete input-output model description will be assumed. For a stable MIMO linear process having a q-dimensional input vector $u(k) = [u_1(k)\ u_2(k)\ ...\ u_q(k)]^T$ and v-dimensional output vector $y(k) = [y_1(k)\ y_2(k)\ ...\ y_v(k)]^T$, the set of unit impulse response matrices H_i, $i = 1...n$, represents the synthetic information describing the process dynamic behavior. The H_i matrix represents the impulse response coefficients at the i-th moment of time ($t = i \cdot T$, where T is the sampling time), having as elements the response of each of the v-outputs to each of the q-unit impulse (Dirac) inputs:

$$H_i = \begin{bmatrix} H_{11}(i) & H_{12}(i) & \cdots & H_{1q}(i) \\ H_{21}(i) & H_{22}(i) & \cdots & H_{2q}(i) \\ \vdots & \vdots & \vdots & \vdots \\ H_{v1}(i) & H_{v2}(i) & \cdots & H_{vq}(i) \end{bmatrix} \tag{2.1}$$

Considering the discrete convolution operation, the output $y(k)$ of the process (at a certain moment of time $t = k \cdot T$) may be computed by:

$$y(k) = \sum_{i=1}^{n} H_i \cdot u(k - i) \tag{2.2}$$

Equation (2.2) is usually denoted as the impulse response model.

For the impulse response model it has been assumed that all process outputs become constant after the time $t = n \cdot T$ (n sampling moments) – that is, a truncated time response is considered. The discrete convolution operation may be determined on the basis of the considered linear property of the process. First, the input function may be described by a sequence (sum) of discrete impulse (Dirac) functions, located at the sampling time moments, each having the impulse area equal to the value of the discrete input signal at the respective sampling time moment. Second, due to linearity, the response of the process to the sequence (sum) of discrete impulse functions is the sum of the individual responses to all previous (time) distributed inputs. As the MIMO case is addressed, each of the v outputs cumulates the responses to all q considered inputs.

Practical considerations make the unit impulse response set of matrices H_i less convenient compared to the unit step response set of matrices S_i, $i = 1...n$. It is well known that the unit step response matrix S_i may be directly related to the unit impulse response matrix H_i by:

$$H_i = S_i - S_{i-1} \tag{2.3}$$

or

$$S_i = \sum_{m=1}^{i} H_m \tag{2.4}$$

Based on the unit impulse–step response relationship, the step response model of the process emerges as:

$$y(k) = \sum_{i=1}^{n-1} S_i \cdot \Delta u(k - i) + S_n \cdot u(k - n) \tag{2.5}$$

where the input change vector $\Delta u(k)$ has been defined by:

$$\Delta u(k) = u(k) - u(k - 1) \tag{2.6}$$

The step response model is again a consequence of the assumed linearity of the process. The input function may be described by a sequence (sum) of step functions, each having the amplitude equal to the value of the discrete input signal at the respective sampling time moment. The response of the process to this sum of step inputs is the sum of the individual step responses to these input steps considered for all transfer paths connecting all inputs to each output and taking into account all previous (but truncated to $n \cdot T$) moments of time.

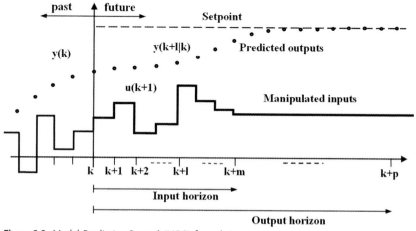

Figure 2.2 Model Predictive Control (MPC) formulation.

Based on the step response model the MPC algorithm may be straightforwardly developed [15]. The MPC approach is simply presented in Figure 2.2.

The present moment is considered to be at $t = k \cdot T$. The future process behavior is considered for a time span ranging from $t = k \cdot T$ to $t = p \cdot T$, the interval usually being denoted as the output horizon. On the basis of the step input model and the current measurements, the process outputs are predicted over the output horizon. The predicted output is a function of the known past, but also of not-yet known future manipulated variables. The future manipulated variables will be computed on a time span ranging from $t = k \cdot T$ to $t = m \cdot T$, with interval usually denoted as the input horizon. The future manipulated variables will be computed such as the predicted outputs conform to a desired (setpoint) output trajectory. The first computed manipulated variable is sent to the process and maintained constant up to the new sampling moment $t = k + 1$, when all the computation is performed again with the output and input horizons shifted one time step forward. This procedure produces the so-called moving or receding horizon control approach, which is one of MPC's characteristics.

Computation of the unknown future manipulated variables, over the input horizon, is performed in an optimization framework, by minimization of the (square) difference between the desired and the predicted output trajectories. The Dynamic Matrix Control formulation uses a linear model for the prediction of the future process behavior and a quadratic performance (programming) QP index (square desired-predicted difference) for determining the unknown manipulated variables. Other linear MPC approaches may consider the linear performance (programming) index or the infinite norm objective function [28]. The basic form of DMC is presented in Eqs. (2.7) to (2.11):

$$
\min_{\Delta u(k)...\Delta u(k+m-1)} \sum_{l=1}^{p} \left(\|y(k+l \mid k) - r(k+l)\|^2_{\Gamma_l^y} + \|\Delta u(k+l-1)\|^2_{\Gamma_l^u} \right) \tag{2.7}
$$

$$
y(k+l \mid k) = \sum_{i=1}^{l} S_i \cdot \Delta u(k+l-i) + \sum_{i=l+1}^{n-1} S_i \cdot \Delta u(k+l-i) + S_n \cdot u(k+l-n) + d(k+l \mid k) \tag{2.8}
$$

$$
d(k+l \mid k) = d(k \mid k) = y_m - \sum_{i=1}^{n-1} S_i \cdot \Delta u(k-i) - S_n \cdot u(k-n) \tag{2.9}
$$

$$
\sum_{l=1}^{p} \left(C_{y_l}^j \cdot y(k+l \mid k) \right) + c^j \leq 0, \quad j = 1 ... n_{c1} \tag{2.10}
$$

$$
\sum_{l=1}^{p} \left(C_{u_l}^j u(k+l-1) \right) + c^j \leq 0, \quad j = 1 ... n_{c2} \tag{2.11}
$$

The notation $y(k+l/k)$ has been used to denote the prediction of the process output vector at the generic time moment $t = k + l$, on the basis of the information known at the present moment $t = k$. The future behavior of the process output is predicted over the prediction horizon, i.e., $l = 1, 2, ..., p$, and consists in the sequence of the predicted process outputs: $y(k+1|k)$, $y(k+2|k)$,..., $y(k+p|k)$.

The notation $d(k+l|k)$ has been used to denote the prediction of the disturbance vector acting on the process output vector at the generic time moment $t = k + l$, on the basis of the information known at the present moment $t = k$. The y_m vector consists in the measured values of the process output vector $y(k)$ at the present moment of time $t = k$.

The notation $||y||_\Gamma^2 = y^T \cdot \Gamma \cdot y$ has been used for the weighted 2-norm of the vector y. The matrix Γ_l^y is the so-called output weighting matrix and is used to penalize the square difference term (between predicted and desired output trajectories) of the optimization index. The matrix Γ_l^u is the so-called input weighting matrix and is used to penalize the manipulated variable move in order to avoid excessive control action. Γ_l^y and Γ_l^u are usually diagonal matrices. $C_{y_l}^j$, $C_{u_l}^j$, c^j are constant matrices of the constraints.

The first term in the optimization index of Eq. (2.7) is the main term from the control point of view as it drives the controlled output to follow the desired setpoint by minimizing the weighted square difference between setpoint and predicted output. The predicted output $y(k+l|k)$ presented in Eq. (2.8) has three main terms.

The first term of Eq. (2.8) corresponds to the component of the output prediction that accounts for the effects of manipulated variables that will be sent to the process in the future, during the input horizon, starting at the present moment of time $t = k \cdot T$ until time $t = m \cdot T$. The values of these future manipulated variables are not yet known, but they are the independent variables of the objective function that will be determined by solving the QP minimization problem. In fact equivalently, the mathematical formulation considers the change of the manipulated variables $\Delta u(k)$ (not the absolute value of the manipulated variables) as the unknown variables.

The second and third terms of Eq. (2.8) correspond to the component of the output prediction that accounts for the past, but known, effects of the (known) manipulated variables that operated in the past $t < k \cdot T$ (starting from the present time $t = k \cdot T$, up to the truncation time $t = (k - n) \cdot T$ in the past).

If the predicted outputs were to consist only of the three terms presented above, and the QP optimization problem were to be solved only on this basis, the resulting control algorithm would be an open-loop optimal control. Consequently, the effects on the output variables of the unmeasured disturbances and of the model–process mismatch would not be taken into consideration, making the control inefficient. Therefore, it is clearly necessary to bring feedback into the control algorithm. This feedback is provided by the last term of Eq. (2.8), which brings the measured values y_m of the output variables onto the scene. In fact, this feedback term is the same as that presented for the internal model control, as MPC is a particular case of the general internal model control structure. The feedback term presented in Eq. (2.9), which is the prediction (estimation) of the disturbances, is usually denoted as the correction component of the predicted output. This correction component may be simply introduced as a constant value, added to the first two components of the prediction all over the prediction horizon, or it may be introduced as a time-varying function if additional information about the disturbance time evolution is available. In its simplest form, the correction term of Eq. (2.9) consists of a constant vector (along the prediction horizon) that is equal to the difference between the measured output at present moment $t = k \cdot T$ and the model-based prediction of the output vector for the same time moment.

The minimization problem is solved at the current moment of time $t = k \cdot T$, but only the first manipulated variable change $\Delta u(k)$ is implemented up until the next sampling time $t = (k + 1) \cdot T$, when the prediction, correction and minimization steps are resumed. This receding horizon approach provides efficiency to the DMC algorithm. For the unconstrained case it may be shown that a linear time-invariant feedback control law may be obtained:

$$\Delta u(k) = K_{DMC} \cdot E(k + 1|k) \tag{2.12}$$

where $E(k+1|k)$ is the vector of future predicted errors for zero future manipulated variable moves.

Equations (2.10) and (2.11) consist of the constraints set for the controlled and manipulated variables. Their presence motivates the success that MPC has gained in applications, despite the increased complexity brought to the minimization problem.

It should be noted that the input m and output p horizons are not necessarily equal. Usually, the output horizon is larger than the input horizon. However, for the time period exceeding the input horizon, until the end of the output horizon, the manipulated variable is considered to remain constant.

To conclude, the following features characterize the MPC algorithm:

- *Prediction*: In contrast to other feedback controllers that compute the control action based on present or past information, MPC determines the control action based on the predicted future dynamics of the system being controlled. The model used for prediction can be either linear or nonlinear, time-continuous or discrete, deterministic or stochastic. Due to the future prediction, early control action can be taken, accounting for the future behavior.
- *Constraints*: In practice, most systems have to satisfy input, state or output constraints, thereby placing limitations on the achievable control performance (in the extreme case affecting stability). One of MPC's incentives is the fact that constraints on the controlled and manipulated variables can be specified in a direct manner. The MPC algorithm is able to find the best (from an optimal objective viewpoint) solution to satisfy all constraints. However, for the general case the controller becomes nonlinear and time-dependent, and no simple stability analysis is available when constraints are active. MPC is able to obtain better control performance, as it can determine the current control action for minimizing errors caused by reaching the constraints that are predicted to become active in the future. Input constraints typically reflect limits on the capacity of control actuators, such as valves or pumps, whereas state constraints represent desirable ranges of operation for process variables, such as temperatures or concentrations. Constraints, however, limit the set of initial conditions,

starting from where a process can be stabilized at a possibly open-loop unstable steady state. Therefore, in the control of constrained processes, it is important to obtain an explicit characterization of the region of closed-loop stability. MPC provides a suitable framework for implementing control that respects manipulated input and process variable constraints, while meeting prescribed performance objectives. Unfortunately, the implicit nature of the feedback law in MPC (the control action is computed by solving on-line a constrained optimization problem at each sampling time) makes the a priori computation of the closed-loop stability region a very difficult task. However, such a computation is possible when Lyapunov-based bounded control techniques are used to design controllers for the stabilization of systems with manipulated input constraints [38].

- *Optimization*: The objective function specifying the desired control performance is minimized on-line at each time step. This objective function is usually an integral weighted square error between predicted controlled variables and desired references.
- *Discrete manipulated variables*: The number of computed values in the manipulated variable sequence is finite (finite input horizon) and discrete in time. They are related to the fact that the involved optimization problem can only be solved with numerical methods. A time-continuous approach can lead to numerically extremely demanding problems.
- *Multivariable character*: Multivariable controllers are often the only solution able to provide the desired control performance in the presence of interactions, though MPC can directly handle such cases by using MIMO models.

2.3
Nonlinear Model Predictive Control

2.3.1
Introduction

Beginning during the past decade, nonlinear model predictive control (NMPC) techniques have become increasingly important and now accepted in chemical industries. The NMPC paradigm encompasses a significant number of different approaches, each with its own special feature. All NMPC techniques rely on the concept of generating values for process inputs as solutions of an on-line (real-time) optimization problem using a nonlinear process model. The nonlinear dynamic model can be used in different ways and in different phases of the control algorithm, depending on the particular nonlinear model predictive control approach. Excellent reviews have been published recently, indicating the current and continuously increasing interest directed toward this control strategy [39–48]. Here, we present a brief overview of the technology, the intention being to demonstrate the diversity of the approaches. While this section represents a quite extensive review of NMPC, the reader is referred to the aforementioned reviews (and the references therein) for a comprehensive overview of the subject.

Among the wide variety of chemical processes encountered, nonlinearity is the rule rather than the exception. There are processes which present many challenging control problems, including: nonlinear dynamic behavior; multivariable interactions between manipulated and controlled variables; unmeasured state variables; unmeasured and frequent disturbances; high-order and distributed processes; and uncertain and (variable) deadtime on inputs and measurements. Further, reliable measurements of important variables to be controlled, such as quality-related variables, are often difficult to obtain on-line. Today, the economic benefits of applying advanced process control (APC) approaches in chemical industry has been widely recognized, not only in academia but also in industry. The theoretical economical optimal operating condition of a chemical process usually lies on active constraints. Therefore, in practice the operating region must be chosen such that the constraints are not violated, even in the case of strong disturbances. The quality of control determines how close the process can be pushed to the boundary. APC approaches allow the tighter control of process variables, hence permitting processes to be operated closer to the limits, and yielding greater profit. A simple graphical explanation of the economical advantages of APC is shown in Figure 2.3. The schematic representation in Figure 2.4 shows that the optimal operating region given by APC is usually on active constraints, but provides higher quality with lower variability than typical operating regions with classical control approaches.

A number of APC approaches and algorithms capable of handling some of the aforementioned process characteristics have been presented during recent years. Many of these approaches are unable to handle the various process characteristics and requirements met in industrial applications, and this has resulted in a large gap between the number of industrial and academic NMPC products. Although, it is well recognized that the performance of a control system is mostly inherent in how successfully it can cope with the *nonlinearity of the process*, chemical processes have been traditionally controlled by using algorithms based on a linear time-invariant approximate process model, the most common being step and impulse response models derived from the convolution integral. One reason for this is that in most LMPC applications reported, the goal is largely to maintain the process at a

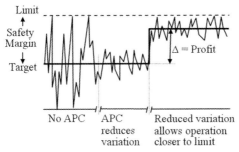

Figure 2.3 Economical benefit given by the use of advanced process control (APC).

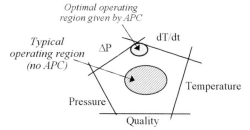

Figure 2.4 Typical operating regions with and without APC.

desired steady state (regulator problem, disturbance rejection; e.g., in refinery processing), rather than moving rapidly from one operating point to another (setpoint tracking problem). A carefully identified linear model, which usually can be identified in a fairly straightforward manner from process test data, is sufficiently accurate in the neighborhood of a single operating point. In addition, by using a linear model and a quadratic objective, the nominal MPC algorithm takes the form of a highly structured convex Quadratic Program (QP), for which reliable solution algorithms and software can easily be found. This is important because the solution algorithm must converge reliably to the optimum in no more than a few tens of seconds if it is to be useful in manufacturing applications. For these reasons, a linear model will in many cases provide the majority of the benefits possible with MPC technology [49–51].

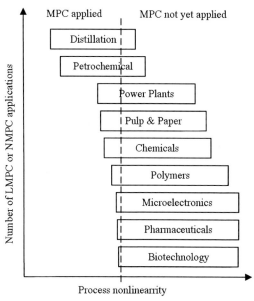

Figure 2.5 Distribution of MPC application versus the degree of process nonlinearity.

Nevertheless, there are cases where nonlinear effects are significant enough to justify the use of NMPC technology. These include at least two broad categories of applications [47]: (i) disturbance rejection control problems where the process is highly nonlinear and subject to large frequent disturbances (pH control, bioreactor, etc.) [52–54]; or (ii) setpoint tracking problems where the operating points change frequently and incorporate a sufficiently wide range of nonlinear process dynamics (batch process control, start-up problems, polymer manufacturing, etc.) [55–58].

An approximate distribution of the number of MPC applications versus the degree of process nonlinearity is shown in Figure 2.5. MPC technology has not yet penetrated deeply into areas where process nonlinearities are strong, and market demands require frequent changes in operating conditions. Consequently, it is these areas that provide the greatest opportunity for NMPC applications [47,50].

A direct extension of the LMPC methods results when a nonlinear dynamic process model is used, rather than the linear convolution model. The nonlinear dynamic model can be used in different ways and in different phases of the control algorithm, depending on the certain NMPC approach [54,57].

2.3.2
Industrial Model-Based Control: Current Status and Challenges

Because of all the appealing features mentioned in the previous section, LMPC has during the past 20–30 years – become a preferred control strategy for a large number of processes. While the development of LMPC approaches and industrial products has reached the fourth generation [59], with more then 4500 applications (Fig. 2.6), there are only few NMPC providers (Table 2.1) with rather limited number of applications. By the late 1970 s, DMC and IDCOM had been developed which basically are at the basis of most of the current LMPC products. The vast majority of LMPC products use a collection of SISO step response models. A fourth-generation LMPC product not mentioned in Qin's review is that of Predict&Control, which was developed by ABB. ABB's approach is based on MIMO state-space models, making it applicable for unstable processes. Process input and output disturbances can be estimated using the incorporated Kalman filter ap-

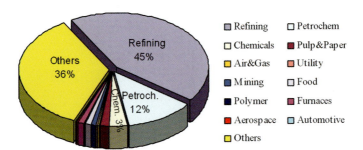

Figure 2.6 Summary of industrial linear Model Predictive Control (LMPC) application by area.

proach. The optimization problem is formulated as a multi-objective quadratic optimization problem with prioritized control objectives and time-domain tuning constraints.

Several commercial NMPC products, together with details of the supplying companies, are listed in Table 2.1 (this list is by no means exhaustive). Several other approaches are available, but these are either mainly applications to a particular process, or the NMPC products are too new (Fischer-Rosemount, ABB, Cybernetica, Ipcos, etc.) and were not included here due to lack of information. One example would be ABB's NMPC product which was recently applied successfully to an industrial drum boiler [60,61]; alternatively, Cybernetica provides a solution to challenge batch NMPC development. Although not all products are included in the review, the technology sold by the companies in Table 2.1 is representative of the current state of the art [59,62].

Table 2.1 NMPC companies and product names.

Company	Product name (Acronym)
Adersa	Predictive Functional Control (PFC)
Aspen Technology	Aspen Target
Continental Controls	Multivariable Control (MVC)
DOT Products	NOVA Nonlinear Controller (NOVA-NLC)
Pavilion Technologies	Process Perfect
Cybernetica	Batch NMPC
IPCOS	INCA

Excellent reviews and descriptions of these NMPC products have been provided [47,59]. Information on the details of each algorithm, including the model types used, options at each step in the control calculation, and the optimization algorithm used to compute the solution, is provided in Table 2.2.

Table 2.2 Comparison of industrial NMPC control technology
(from [177]).

Company	Adersa	Aspen Technology	Continental Controls	DOT Products	Pavilion Technologies
Algorithm	PFC	Aspen Target	MVC	NOVA-NLC	Process Perfect
Model Forms[1]	NSS-FP, S, I, U	NSS, S, I, U	SNP-IO, S, I	NSS-FP, S, I	NNN-IO, S, I, U
Feedback[2]	CD, ID	CD, ID, EKF	CD	EKF	CD
SS. Opt. Obj.[3]	Q [I,O]	Q [I,O]	Q [I,O]	-	Q [I,O]
SS. Opt. Const.[4]	IH, OH	IH, OH	IH, OS	-	IH OH
Dyn. Opt. Obj.[5]	Q [I,O]	Q [I,O,M]	Q [I,O,M]	(Q,A) [I,O,M]	Q [I,O]
Dyn. Opt. Const.[6]	IC, OS	IH, OH, OS	IH, OS	IH, OS	IH, OH
Output Traj.[7]	S, Z, RT	S, Z, RT	S, Z, RT	S, Z, RT	S, TW
Output Horiz.[8]	FH, CP	FH, CP	FH	FH	FH
Input Param.[9]	BF, SM	MM	SM	MM	MM
Sol. Method[10]	NLS	QP, QP-QUICK	GRG, GRG2	MINLP-Nova	GD

[1] Model Form: IO = Input-Output; FP = First-Principles; NSS = Nonlinear State-Space; NNN = Nonlinear Neural Net; SNP = Static Nonlinear Polynomial; S = Stable; I = Integrating; U = Unstable.

[2] Feedback: CD = Constant Output Disturbance; ID = Integrating Output Disturbance; EKF = Extended Kalman Filter.

[3] Steady-State Optimization Objective: Q = Quadratic; I = Inputs; O = Outputs.

[4] Steady-State Optimization Constraints: IH = Input Hard maximum, minimum, and rate of change constraints; OH = Output Hard maximum and minimum constraints.

[5] Dynamic Optimization Objective: Q = Quadratic; A = One norm; I = Inputs; O = Outputs; M = Input Moves.

[6] Dynamic Optimization Constraints: IH = Input Hard maximum, minimum, and rate of change constraints; IC = Input Clipped maximum, minimum, and rate of change constraints; OH = Output Hard maximum and minimum constraints; OS = Output Soft maximum and minimum constraints.

[7] Output Trajectory: S = Setpoint; Z = Zone; RT = Reference Trajectory; TW = Trajectory Weighting.

[8] Output Horizon: FH = Finite Horizon; CP = Coincidence Points.

[9] Input Parameterization: SM = Single Move; MM = Multiple Move; BF = Basis Functions.

[10] Solution Method: NLS = Nonlinear Least Squares; QP = Quadratic Program; GRG = Generalized Reduced Gradient; GD = Gradient Descent; MINLP = Mixed Integer Nonlinear Program.

An approximate summary of industrial NMPC applications, together with a breakdown of the different areas of application, is shown in Figure 2.7.

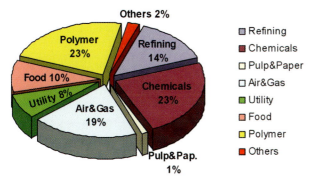

Figure 2.7 Summary of nonlinear model predictive control (NMPC) applications by area.

It is of interest to note that the recent distribution of NMPC applications has changed significantly compared to LMPC applications. In refining, where processes are operated around a steady-state, LMPC is effective and the additional burden related to NMPC is probably not necessary, but for polymers and chemicals (e.g., pharmaceuticals) NMPC appears to offer the best approach. In fact, while half of the NMPC applications are in these two areas, only a very small percentage of LMPC applications are used. The explanation for this might be due to the fact that these industries are dominated by highly nonlinear and unstable batch processes, where LMPC approaches usually fail. An analysis in these areas showed that these applications are for continuous processes where LMPC approaches (sometimes with gain scheduling) can function around the operating points. For example, INCA® (the product of IPCOS Technology [63]) in the actual control calculations uses linear models carefully identified for the operating points, but this approach is more difficult to use efficiently for batch processes. Likewise, for polymers and chemicals (e.g., pharmaceuticals), the objectives are in terms of distribution control of the final product quality, which is difficult to express in the LMPC framework. An interesting observation from the analysis of the NMPC applications showed that almost all literature-reported applications were for continuous processes, and that many of the presented products were not produced for discontinuous highly nonlinear and unstable processes with wide variations in process gain. The need for NMPC in continuous polymerization processes can be explained by the frequent grade changeover operations of the reactors in order to meet the diversified demands of the market. However, even in these cases a relatively simpler NMPC approach based on linear model scheduling can be successful, and may be applied. In the case of batch polymerization and certain fine chemical processes, the physico-chemical properties of the system (viscosity, density, heat capacity) can alter dramatically during the process, leading to changes in the gain – sometimes of two orders of magnitudes. In these cases, the controller must cover a wide range of operating conditions and cope with highly nonlinear process dynamics.

2.3.2.1 Challenges in Industrial NMPC

The flow of tasks to be performed in generic industrial NMPC applications is presented schematically in Figure 2.8. Each block represents a specific problem which must be discussed in more detail, and can represent the main framework of industrial NMPC assessments. Even the link of the NMPC to the process through the input/output devices (I/O) is an important issue, where communication protocols between process and controller – or even between different task-blocks within the NMPC – need to be compared (OPC, DDE, UDP, TCP/IP, etc.). Clearly, not all blocks are present for all NMPC applications. However if the goal is to develop generic NMPC tools adequate for SISO, or thin (CVs > MVs), fat (CVs < MVs), or square (CVs = MVs) MIMO plants, then all components must be considered, and those not needed for a certain application will be turned off. For example, in the case of SISO control systems the determination of process subsets and ill-conditioning is not an issue, and the corresponding modules will be bypassed in the controller.

Some of the major challenges related to industrial NMPC applications include the following.

2.3.2.1.1 Efficient Development and Identification of the Control-Relevant Model

The importance of modeling in NMPC applications is well acknowledged. Unlike traditional control, where modeling and analysis represent a small part of the effort in the application development, it is estimated that up to 80% of time and expense in the design and installation of a NMPC is attributed to modeling and system identification. The design effort involved in NMPC design versus traditional control is shown schematically in Figure 2.9.

All LMPC (and most NMPC) products use empirical models identified through plant tests. It is attractive to use first-principles (FP) models, which provide the most

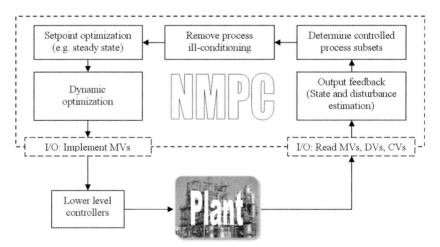

Figure 2.8 Sequence of tasks involved in a generic NMPC application.

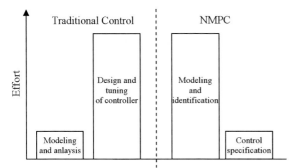

Figure 2.9 Design effort of NMPC versus traditional control.

inside information about the process. FP models are globally valid and therefore well suited for optimization that can require extrapolation beyond the range of data used to fit the model. Despite the clear advantages of using FP models, most NMPC approaches are based on empirical nonlinear models. The reasons for this relate to the difficulty in developing proper process models, and in the increased computational burden required to solve FP models. Since the modeling is a time-consuming part of the NMPC design, the choice of a correct modeling environment is crucial. Almost all industrial NMPC approaches attempt to benefit from the power of chemical modeling software (ASPEN, gPROMS, Hysys, etc.) rather than building and solving the models from scratch. The model identification is important, even if FP models are used, and in this case offline parameter identification has usually to be performed. It is very important to bear in mind (however often overlooked) that models (whether empirical or FP) are imperfect (both in terms of structure and parameters); therefore, robustness is a key issue in NMPC applications. Robustness has also been identified as one of the major deficiencies of current NMPC products (actually none of the products presented here has any systematic treatment of robust performance except the inherent robustness due to feedback, and penalization of excessive control movements). How to choose the correct plant tests to identify the best model is an important question [64,65]. For example, many vendors believe that the plant test is the single most important phase in the implementation of DMC-plus controllers. *Optimal experimental design*, even if not *the* optimal design, can provide an answer/recipe to the model identification question. Additionally, robust NMPC design can lead to significant performance improvement [66,67]. Another important problem which must be assessed in the case of FP model-based NMPC approaches, is the trade-off between the model accuracy (hence complexity) and computational requirement. Often, the control-relevant model is not necessarily the most accurate, and identifying the correct balance between the accuracy of the model and computational burden is usually very challenging. The inclusion of many details into the model can lead to a large number of states, resulting in unobservable models based on available measurements, in addition to a prohibitively large computational burden. Determination of the control-relevant model must always be carried out in conjunction with the observer design.

2.3.2.1.2   State and Parameter Estimation

The lack of reliable *sensors* is one of the major bottlenecks in industrial NMPC applications; this problem is crucial when FP models are used [68]. In many industrial applications software sensors must be developed based on additional empirical models (most often neural networks are used for this purpose, e.g., INCA, [63]).

2.3.2.1.3   Managing the On-Line Computational Load

Although an exorbitant computational load is involved in NMPC approaches, the use of an efficient optimization approach is crucial [69–73]. Other important problems that are related to computations involved in NMPC, and which must be assessed for successful practical implementation, include the robustness of optimization, the choice of proper objective functions, and the optimization problem set-up for efficient solution.

2.3.2.1.4   Design as an Integrated System

In industrial NMPC, optimization is involved in several steps of the controller design (steady-state optimization, dynamic optimization, state estimation, and model identification). An efficient approach to design model, estimator (of model parameters and states), and optimization algorithms as an integrated system (simultaneously optimized) rather than as independent components should be the capstone of modern NMPC designs.

2.3.2.1.5   Long-Term Maintenance of Control System

Although it is clear that the implementation complexity of NMPC approaches is high, it is important to asses how long-term maintenance can be performed, and also to identify the limits of the approach in the face of changing process and operating conditions.

2.3.3
First Principle (Analytical) Model-Based NMPC

The objective of NMPC is to calculate a set of future control moves (control horizon, T_c, or M) by minimization of a cost function, such as the squared control error on a moving finite horizon (prediction horizon, T_P or P). The optimization problem is solved on-line based on predictions obtained from a nonlinear model. Here, it is possible to use different empirical nonlinear models for predictions in the controller, but the most attractive approach is to use FP models. These are globally valid and therefore well-suited to an optimization which might require extrapolation beyond the range of data used to fit the model [74,75].

A general mathematical formulation of the NMPC problem, when the process is described by ordinary differential equations (ODEs), is:

- Objective function:

$$\min_{u(\cdot), P, M} \{ J(x(t), u(\cdot), P, M) \} \tag{2.13}$$

- Constraints:

$$\frac{dx}{dt} = f(x, u, q, d) \tag{2.14}$$

$$x(k) = x_{est}(k) \tag{2.15}$$

$$0 = g_1(x) \tag{2.16}$$

$$y_p = g_2(x) \tag{2.17}$$

$$u_{min}(k + i) \leq u(k + i) \leq u_{max}(k + i) \tag{2.18}$$

$$u(k + i - 1) - \Delta u_{max} \leq u(k + i) \leq u(k + i - 1) + \Delta u_{max} \tag{2.19}$$

$$u(k + i) = u(k + M - 1) \text{for all} i = \overline{M - 1, P} \tag{2.20}$$

$$x_{min}(k + i) \leq x(k + i) \leq x_{max}(k + i) \tag{2.21}$$

$$y_{min}(k + i) \leq y_p(k + i) \leq y_{max}(k + i) \tag{2.22}$$

where x = state variables, u = manipulated variables, q = parameters, d = measured and unmeasured disturbances, and y = output variables.

Generally, the prediction and control horizon, respectively, are considered fixed for an open-loop optimization. The objective function usually is chosen as the sum of the squares of the differences between the predicted outputs and the setpoint values over the prediction horizon of P time steps:

$$J(x(t), u(\cdot)) = \int_{t_k}^{t_k + T_P} \|Q \cdot E\|^2 dt \Leftrightarrow J(x(k), u(\cdot)) = \sum_{i=1}^{P} \|Q_i(r(k + i) - y_p(k + i))\|^2$$

$$\text{continuous form} \qquad \text{discrete form} \tag{2.23}$$

Often, the objective function [Eq. (2.13)] includes a second term, which is the squared sum of the manipulated variable changes over the control horizon (M):

$$J(x(t), u(\cdot)) = \sum_{i=1}^{P} \|Q_i(r(k + i) - y_p(k + i))\|^2 + \sum_{i=1}^{M} \|R_i \Delta u(k + i - 1)\|^2 \tag{2.24}$$

The second term was originally introduced by the unconstrained formulation of LMPC in which constraints are handled artificially through the weighting factors (matrices Q_i and R_i). Since the constrained LMPC algorithms (likewise the NMPC methods) explicitly include constraints, there is no need for the second term in the objective function, although in many constrained LMPC and NMPC applications the authors also use this term. Instead of – or in addition to – the second term of the objective function, a smooth control action can be assured by introducing another term to minimize the deviations of the manipulated inputs from their set points. In this way, a more general formulation of the performance function is obtained and the optimization problem to be solved at each sampling time can be written as follows:

$$
\min_{u(k)...u(k+M-1)} \left\{ \sum_{i=1}^{P} \left\| Q_i \big(r(k+i) - y_p(k+i) \big) \right\|^2 + \sum_{i=1}^{M} \left\| R_i \Delta u(k+i-1) \right\|^2 + \right.
$$
$$
\left. \sum_{i=1}^{M} \left\| R_l \big(u(k+i-1) - u^{ref}(k+i-1) \big) \right\|^2 \right\}
\tag{2.25}
$$

The predicted values of the output variables (y_p) can be considered equal to the values obtained from the model (y_m), but usually a correction is made to reduce the cumulative error effect of the measurement errors and the model/plant mismatch. The correction equation usually has the following form:

$$
y_p(k+i) = y_m(k+i) + K(k+i) \cdot \big(y_m(k) - y_p(k) \big)
\tag{2.26}
$$

The decision variables in the optimization problem expressed by Eqs. (2.13) to (2.22) are the control actions, M sampling time steps into the future (control horizon). Generally, $1 \le M \le P$, and it is assumed that manipulated variables are constant beyond the control horizon [Eq. (2.20)]. Although the optimization provides a profile of the manipulated input moves over a control horizon (M), only the first control action is implemented. After the first control action is implemented, new measurements are obtained and are used for the compensation of plant/model mismatch and for the estimation of unmeasured state variables. Finally, the prediction horizon is shifted by one sampling time into the future and the optimization is performed again.

In the NMPC approaches, absolute [Eq. (2.18)] and velocity [Eq. (2.19)] constraints for manipulated variables, as well as state- and output-variable constraints, are explicitly included.

Since a constrained nonconvex nonlinear optimization problem must be solved on-line, the major practical challenge associated with NMPC is the computational complexity that increases significantly with the complexity of the models used in the controller. Significant progress has been made in the field of dynamic process optimization, with rapid on-line optimization algorithms which exploit the specific structure of optimization problems arising in NMPC having been developed. Moreover, real-time applications have been proven to be feasible for small-scale processes. Nonetheless, the global solution for optimization cannot be guaranteed, and the development of rapid and stable optimization techniques remains a major objective in NMPC research [76].

2.3.4
NMPC with Guaranteed Stability

In addition to the constraints described in Section 2.3.3, another important requirement that the nonlinear model predictive controllers must meet (however seldom it
is taken into consideration for practical implementation) is that it should assure a
stable closed-loop system.

The most straightforward way to achieve guaranteed stability is to use an infinite
horizon cost functional ($P = M = \infty$) [77,78]. In this case, Bellman's principle of
optimality implies that the open-loop input and state trajectories obtained as the
solution of the optimization problem are equal to the closed-loop trajectories of the
nonlinear system. Consequently, any feasible predicted trajectory goes to the
origin. [In this section we consider that the reference values for the controlled
and manipulated variables, respectively, are their steady states values ($r = x_s$, $u^{ref} =
u_s$). Furthermore, without loss of generality, it can be assumed that the origin ($x_s =
0$, $u_s = 0$) is the steady-state point of interest of the system.] However, using infinite
horizon in the performance criterion leads to a practically unsolvable optimization
problem. In order to cope with this disadvantage, and to guarantee stability, so-
called stability constraints (besides the input and state constraints) must be
included into the finite horizon open-loop optimization problem [79–83].

The most widely suggested stability constraint is the terminal equality constraint, which forces the states to be zero (equal to their steady-state values) at
the end of the finite horizon:

$$x(t + T_P) = 0 \tag{2.27}$$

Although using the terminal equality constraint to guarantee stability is an intuitive approach, it also increases significantly the on-line computation necessary
to solve the open-loop optimization problem, and often causes feasibility problems
[84–86].

Another approach to guarantee stability is the so-called *quasi-infinite horizon
nonlinear MPC* (QIHNMPC), in which the prediction horizon is approximately
extended to infinity by introducing a terminal penalty term in the objective
function [87–92]. The basic idea of this approach consists of an approximation of
the infinite horizon prediction to achieve closed-loop stability, while the input
function to be determined on-line is of finite horizon only. The terminal penalty
term is determined off-line, such that it bounds from above the infinite horizon
objective function of the nonlinear system controlled by a local state feedback law
in a terminal region Ω [91,93–97].

The difference between this approach and that described above is that, rather
than using terminal equality constraints, a terminal region is used and the
objective function has an additional term, the terminal penalty term. The key
problem here is the choice of the form, and the off-line computation of the terminal
region and penalty term which, in general, and due to nonlinearity of the system, is
a difficult task. When using a local linear feedback law and a quadratic objective

function, the terminal penalty term can be chosen to be quadratic. In this case, the terminal penalty matrix Q_p is the solution of a Lyapunov equation, and the on-line control problem can be expressed as below:

$$\min_{\bar{u}(\cdot)}\{J(x(t),\bar{u}(\cdot))\} \tag{2.28}$$

with

$$J(x(t),\bar{u}(\cdot)) \overset{def}{=} \int_{t}^{t+T_p} \left(\|\bar{x}(\tau;x(t),t)\|_Q^2 + \|u(\tau)\|_R^2 \right) d\tau + \|\bar{x}(t+T_p;x(t),t)\|_{Q_p}^2 \tag{2.29}$$

subject to:

$$\frac{d\bar{x}}{dt} = f(\bar{x},\bar{u}), \text{with initial condition} :\bar{x}(t;x(t),t) = x(t) \tag{2.30}$$

$$\bar{u}(\tau) \in U, \tau \in \left[t,t+T_p\right] \tag{2.31}$$

$$\bar{x}(\tau;x(t),t) \in X, \tau \in \left[t,t+T_p\right] \tag{2.32}$$

$$\bar{x}(t+T_p;x(t),t) \in \Omega \tag{2.33}$$

where, $\|x\|_Q^2 = x^T \cdot Q \cdot x$, with Q a positive-definite matrix, and the bar indicating that the corresponding variables are predicted values. In this case, the procedure to obtain the terminal penalty matrix and the terminal region is presented below [90,98]:

- Step 1. Consider the Jacobian linearization of system (2.30) at the origin:

$$\frac{dx}{dt} = A \cdot x + B \cdot u \tag{2.34}$$

where:

$$A = \left.\frac{\partial f}{\partial x}\right|_{(0,0)} \text{ and } B = \left.\frac{\partial f}{\partial u}\right|_{(0,0)} \tag{2.35}$$

- Step 2. Find a locally stabilizing linear state feedback $u = Kx$, by solving the control problem based on the Jacobian linearization. Introducing $u = Kx$ in Eq. (2.34) one obtains:

$$\frac{dx}{dt} = A \cdot x + B \cdot u = A \cdot x + B \cdot K \cdot u = (A + B \cdot K) \cdot x = A_K \cdot x \tag{2.36}$$

If Eq. (2.34) can be stabilized, then $A_K \overset{def}{=} A + B \cdot K$ is asymptotically stable, and the Lyapunov Eq. (2.37) has a unique positive-definite and symmetric solution, Q_p.
- Step 3. Choose a constant $\varkappa \in [0,\infty)$, with $\varkappa < -\lambda_{max}(A_K)$, where $\lambda_{max}(A_K)$ denotes the largest eigenvalue of the matrix A_K, and solve the Lyapunov equation:

$$(A_K + \varkappa \cdot I)^T \cdot Q_p + Q_p \cdot (A_K + \varkappa \cdot I) = -(Q + K^T \cdot R \cdot K) \tag{2.37}$$

- Step 4. Find the region defined by Eq. (2.38) with the largest possible α_1:

$$\Omega_1 \stackrel{def}{=} \{x \in \Re^n | x^T Q_p x \leq \alpha_1\} \tag{2.38}$$

such that $\Omega_1 \subseteq X$ and $Kx \in U$, for all $x \in \Omega_1$.
- Step 5. Find the largest possible terminal region:

$$\Omega \stackrel{def}{=} \{x \in \Re^n | x^T Q_p x \leq \alpha\} \tag{2.39}$$

by determining $\alpha \in (0, \alpha_1]$, via iteration, until the optimal value of the following optimization problem is non-positive:

$$\max_x \{x^T Q_p \Phi(x) - \varkappa \cdot x^T Q_p x | x^T Q_p x \leq \alpha\} \tag{2.40}$$

where, $\Phi(x) \stackrel{def}{=} f(x, Kx) - A_K x$.

2.3.5
Artificial Neural Network (ANN)-Based Nonlinear Model Predictive Control

2.3.5.1 Introduction
The analytical model-based NMPC has two main disadvantages:
- It requires the elaboration of a complex, analytical model of the process with good accuracy which, in the case of the most chemical processes, might be very arduous.
- In order to carry out the optimization problem by integrating the analytical model for large scale, complicated processes may demand much computation effort and time.

These shortcomings can be avoided by using ANNs as the nonlinear model to control movement computation. The advantageous properties of neural networks, such as parallel computation, nonlinear mapping and learning capabilities, make them an alluring tool in many chemical engineering problems. In recent years, there has been intense, growing interest in this field of artificial intelligence [99]. Neural networks have been used successfully for a wide variety of chemical applications, including: the detection and location of gross errors in process systems [100]; the detection of faults in control systems [101]; the optimal design of chemical processes [102–104]; elucidating nonlinear input-output maps for process data; the identification and modeling of linear and nonlinear systems [105–109]; process control [110–113]; pattern recognition [114,115]; and several other purposes [116,117]. In addition to the above-mentioned reports concentrating on the applications of neural networks, there has been a recent increase in the number of studies related to the control-relevant properties of neural networks

[118], the improvement of network training by using different network structures, transfer functions and learning algorithms [109,119–121], as well as the elucidation of definite methodologies to determine network structures [122].

2.3.5.2 Basics of ANNs

A neural network is a computer program architecture for nonlinear computations, which is composed of many simple elements operating in parallel. These elements, termed "processing elements", are inspired by biological nervous systems, and are highly interconnected. An individual processing element (neuron) can have any number of inputs, but only one output that is generally related to the inputs by a transfer function. The most frequently used transfer functions include: sigmoid function; hyperbolic tangent function; sine function; and linear and saturated linear transfer function. The most frequently used transfer functions are listed in Table 2.3. The argument for the transfer function is the sum of the input elements of the corresponding neuron, with each input being multiplied by the associated weight; this demonstrates the strength of the connection between two neurons.

Table 2.3 Transfer functions used in the artificial neurons.

Function name	Expression	Characteristics
Hard limit	$g(x) = \begin{cases} +1, & x > 0 \\ 0, & \text{otherwise} \end{cases}$	Nondifferentiable, step
Symmetric hard limit	$g(x) = \begin{cases} +1, & x > 0 \\ -1, & \text{otherwise} \end{cases}$	Nondifferentiable, step
Log-sigmoid	$g(x) = \dfrac{1}{1 + e^{-x}}$	Nondifferentiable, step, positive mean
Tan-sigmoid	$g(x) = \tanh(x)$	Differentiable, step, zero mean
Radial	$g(x) = e^{-x^2 - a^2}$	Differentiable, impulse, positive mean

A neuron usually has an additional input, called bias, which is much like a weight corresponding to a constant input of 1. A schematic representation of an individual processing element (neuron) is shown in Figure 2.10.

The neurons are typically grouped into subsets, called layers, in which usually all the process units have the same bias and transfer function. Among the various architectures proposed for neural networks, the multi layer, feed-forward network has been used most frequently for dynamic modeling and process control applications. A typical feed-forward neural network has one input layer (usually with the identity transfer function; thus, it only distributes the inputs to the neurons from the next layer), one output layer, and one or more hidden layers. The structure of a feed-forward neural network is shown in Figure 2.11.

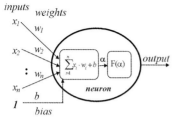

Figure 2.10 An individual processing element of a neural net.

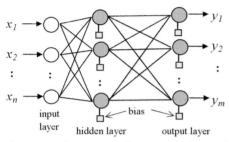

Figure 2.11 The structure of a general feed-forward neural network.

The outputs of the neurons from a layer represent the inputs for the next layer. The architecture of a network consists of a description of how many layers the network contains, the number of neurons in each layer, the transfer function used in each layer, and how the layers are connected to each other.

2.3.5.3 Algorithms for ANN Training

Training the ANN is the most time-consuming task in the ANN application development. It is very difficult to recognize which training algorithm will be the fastest for a given problem, as success will depend on many factors, including the complexity of the problem, the number of data points in the training set, the number of weights and biases in the network, and the error goal.

2.3.5.3.1 Back-Propagation Learning Algorithm

The most frequently used learning algorithm to train feed-forward networks is the back-propagation learning algorithm, due mainly to its simplicity. The main advantage of this algorithm is that it is easily implemented and requires little computational power; thus, it can be used to train large networks. During the training phase, the connection weights and biases are modified, using the back-propagation learning rules, so that the network will learn the process features.

The back-propagation learning algorithms belong to the class of supervised training algorithms; that is, there is a set of input-output data which is repeatedly presented to the network when the weights are adjusted in order to minimize the error between the net output (A_N) and the desired training output (D). The greater the number of sets of input-output data (Q), the better the network will learn the process. Hence, for a network the training data can be represented with the following matrices:

$$P = \begin{bmatrix} P(1,1) & P(1,2) & \cdots & P(1,Q) \\ P(2,1) & P(2,2) & \cdots & P(2,Q) \\ \vdots & \vdots & & \vdots \\ P(S_0,1) & P(S_0,2) & \cdots & P(S_0,Q) \end{bmatrix} \tag{2.41}$$

$$D = \begin{bmatrix} D(1,1) & D(1,2) & \cdots & D(1,Q) \\ D(2,1) & D(2,2) & \cdots & D(2,Q) \\ \vdots & \vdots & & \vdots \\ D(S_N,1) & D(S_N,2) & \cdots & D(S_N,Q) \end{bmatrix} \tag{2.42}$$

The back-propagation algorithm can be represented briefly as follows:

- Initialization of the weight coefficients with random values.
  do
  For each training input-output pair:
  – the input array is presented to the network and the activation flux is propagated layer by layer through the net (forward step).
  – an error criterion is calculated and it is propagated back through the net adjusting the weights in order to minimize the error criterion (backward step).
  while (error is above the error goal).
- Forward step
  In this step, the output of the net is calculated. For a feed-forward network with N layers (the input layer is not counted), with S_j neurons in the jth layer and with the same transfer function (F_j) in one layer, the output of the jth layer can be computed with the following matrix with recurrent terms:

$$A_j = \begin{bmatrix} F_j\left(\sum_{i=1}^{S_{j-1}} w_j(1,i) \cdot A_{j-1}(i,1) + B_j(1)\right) & \cdots & F_j\left(\sum_{i=1}^{S_{j-1}} w_j(1,i) \cdot A_{j-1}(i,Q) + B_j(1)\right) \\ \vdots & \vdots & \vdots \\ F_j\left(\sum_{i=1}^{S_{j-1}} w_j(S_j,i) \cdot A_{j-1}(i,1) + B_j(S_j)\right) & \cdots & F_j\left(\sum_{i=1}^{S_{j-1}} w_j(S_j,i) \cdot A_{j-1}(i,Q) + B_j(S_j)\right) \end{bmatrix} \tag{2.43}$$

with $F_0 = \Im$ (identity function: $\Im(x) = x$) and $A_0 = P$. For $j = N$ the output of the network is obtained.

- Backward step

The most frequently used error criterion, calculated in this step is the sum-squared error of the network, defined as:

$$E = \sum_{i=1}^{S_N} \sum_{j=1}^{Q} (D(i,j) - A_N(i,j))^2 \tag{2.44}$$

The adjustment of the network weights and biases is made by continuously changing their values in the direction of steepest descent with respect to error. There are several improved methods to perform this task. One such algorithm is the back-propagation learning with momentum [123]. Momentum (m) allows the network to ignore shallow local minimums in the error surface, and can be added to back-propagation learning by making weight changes equal to the sum of a fraction of the last weight change and the new change suggested by the back-propagation rule. This is expressed mathematically as:

$$w(i,j)^{(h)} = w(i,j)^{(h-1)} + l_r \cdot \delta w(i,j)^{(h)} \tag{2.45}$$

where

$$\delta w(i,j)^{(h)} = m \cdot \delta w(i,j)^{(h-1)} + (1-m) \cdot \frac{\partial E}{\partial w(i,j)} \tag{2.46}$$

These two steps (forward and backward) are repeated until the sum squared error (E) becomes less than the error goal.

2.3.5.3.2 The Levenberg–Marquardt Algorithm

In general, for networks which contain up to a few hundred weights, the Levenberg–Marquardt algorithm will have the fastest convergence. This advantage is especially noticeable if very accurate training is required.

The Levenberg–Marquardt algorithm was designed to approach second-order training speed without having to compute the Hessian matrix. When the performance function has the form of a sum of squares (as it is typically in training feedforward networks), then the Hessian matrix can be approximated as:

$$\mathbf{H} = \mathbf{J}^T \mathbf{J} \tag{2.47}$$

and the gradient can be computed as:

$$\mathbf{g} = \mathbf{J}^T \mathbf{e} \tag{2.48}$$

where $\mathbf{J}$ is the Jacobian matrix, which contains first derivatives of the network errors with respect to the weights and biases, and $\mathbf{e}$ is a vector of network errors. The Jacobian matrix can be computed through a standard back-propagation technique that is much less complex than computing the Hessian matrix.

The Levenberg–Marquardt algorithm uses this approximation to the Hessian matrix in the following update:

$$\mathbf{x}_{k+1} = \mathbf{x}_k - \left[\mathbf{J}^T\mathbf{J} + \mu\mathbf{I} \right]^{-1} \mathbf{J}^T\mathbf{e} \tag{2.49}$$

When the scalar μ is zero, this is simply Newton's method, using the approximate Hessian matrix. However, when μ is large, this becomes a gradient descent with a small step size. Newton's method is faster and more accurate near an error minimum, so the aim is to shift towards Newton's method as quickly as possible. Thus, μ is decreased after each successful step (reduction in performance function) and is increased only when a tentative step would increase the performance function. In this way, the performance function will always be reduced at each iteration of the algorithm.

2.3.5.3.3 Using Bayesian Regularization to Obtain the Best ANN Model

One problem that occurs during neural network training is termed "overfitting". The error on the training set is driven to a very small value, but when new data are presented to the network the error is large. The network has memorized the training examples, but it has not learned to generalize to new situations. One method for improving network generalization is to use a network, which is just large enough to provide an adequate fit. The larger the network used, the more complex the functions that the network can create. If a small enough network is used, it will not have enough power to overfit the data. The problem is that it is difficult to know beforehand how large a network should be for a specific application.

Among the ANN-based modeling society, much important effort has been invested in the development of different algorithms, which can help to avoid the phenomena of overfitting. Generally speaking, there are two ways to obtain the ANN model with the best generalization properties [124]:

- Using special algorithms to determine the optimal topology of the ANN model (Optimal Brain Damage, Optimal Brain Surgeon, etc.); this approach is described in detail in Chapter 3.
- The second approach is based on introducing different modifications into the training algorithms, so that the resulting network will not overfit the data (early stopping, regularization, etc.).

One of the most advanced algorithms from the second approach is the so-called "regularization". This algorithm involves modifying the performance function, which is normally chosen to be the sum of squares of the network errors on the training set. It is possible to improve generalization if the performance function is modified by adding a term that consists of the mean of the sum of squares of the network weights and biases:

$$F = \gamma \cdot \frac{1}{S_N} \cdot \sum_{i=1}^{S_N} \left(\frac{1}{Q} \cdot \sum_{j=1}^{Q} (A_i(j) - T_i(j))^2 \right) + (1 - \gamma) \cdot \frac{1}{N_p} \cdot \sum_{k=1}^{N_p} w_k^2 \tag{2.50}$$

where γ is the performance ratio.

Using this performance function will cause the network to have smaller weights and biases, and this will force the network response to be smoother and less likely to overfit. The problem with regularization is that it is difficult to determine the optimum value for the performance ratio parameter. If this parameter is made too large, there may be overfitting, whereas if the ratio is too small, the network will not adequately fit the training data. It is desirable to determine the optimal regularization parameters in an automated fashion. One approach to this process is the Bayesian framework of David MacKay [121]. In this framework, the weights and biases of the network are assumed to be random variables with specified distributions. The regularization parameters are related to the unknown variances associated with these distributions. These parameters can then be estimated using statistical techniques. A detailed discussion of the use of Bayesian regularization, in combination with Levenberg–Marquardt training, can be found in [125].

Besides the aforementioned algorithms, there is an important number of training algorithms. A new improved training algorithm, as elaborated by the present authors, is presented in [126]. This algorithm combines the advantages of genetic algorithms and the Levenberg–Marquardt algorithm, and is used to obtain the ANN model of a fluid catalytic cracking unit.

2.3.5.4 Direct ANN Model-Based NMPC (DANMPC)

There are several ways to use ANNs in control. In the description of the control algorithms based on ANN models, it is considered that the process model can be written in a discrete form [99,127]:

$$y(k+1) = f(y(k), \ldots, y(k-n+1), u(k), \ldots u(k-m+1)) \tag{2.51}$$

where $u(k)$ and $y(k)$, $k = 0, 1, \ldots$ are input and output vectors at the time instances k, n and m are the orders of $\{y(k)\}$ and $\{u(k)\}$ respectively, and f is a nonlinear function which is thought to be unknown, though some idea of its structure is probably apparent. Considering the correct time-delay of the process might be crucial for obtaining the appropriate ANN model. The time-delay (d; $d \geq 0$) of the process can be introduced in the ANN model in the following way:

$$y(k+1) = f(y(k), \ldots, y(k-n+1), u(k-d), \ldots u(k-d-m+1)) \tag{2.52}$$

The order of the process outputs (n) as well as the time-delay (d) must be known. Usually, these values can be estimated from experience.

The DANMPC relies on the consideration of an approximate function of f expressed by an ANN. For this, it is considered that $y(k)$, $\ldots$, $y(k-n+1)$, $u(k-d)$, $\ldots$, $u(k-d-m+1)$ and $y(k+1)$ are the inputs and the outputs of the network respectively, and an approximate dynamic model is constructed by adjusting a set of connection weights and biases (W) via training using historical data:

$$\hat{y}(k+1) = f_{ANN}(y(k), \ldots, y(k-n+1), u(k-d), \ldots u(k-d-m+1); W) \tag{2.53}$$

This network can be used in a *successive recursive* way in a general NMPC structure, to obtain the model prediction. This model can be used for prediction in several ways. Below are presented two types of *d*-step ahead predictors to compensate the influence of the time-delay.

2.3.5.4.1 Recursive *d*-Step Ahead Predictor

Based on the model in Eq. (2.53), one can use a successive recursive technique to obtain the *d*-step ahead predictor expressed below:

$$\hat{y}(k+d+1) = f_{ANN}(\hat{y}(k+d), \ldots, \hat{y}(k+d-n+1), u(k), \ldots, u(k-m+1); \ W) \tag{2.54}$$

This predictor depends on the predictions at the previous steps within the prediction horizon. The recursive property of the predictor enables the extension for long-range prediction. Suppose the prediction horizon is *P*, the long-range predictor can be described as:

$$\hat{y}(k+i|k) = f_{ANN}(\hat{y}(k+i-1), \ldots, \hat{y}(k+i-\min(i,n)), y(k), \ldots, y(k-\max(n-i,0)), u(k+i-d), \ldots, u(k+i-d-m); \ W); \quad 1 \le i \le P \tag{2.55}$$

where $\hat{y}(k+i|k)$ is the predicted value of *y* for the moment *k+i* obtained at moment *k* based on the information available up to moment *k*.

2.3.5.4.2 Non-Recursive *d*-Step Ahead Predictor

The non-recursive *d*-step ahead predictor uses the sequences of both past inputs and outputs of the process until sampling time *k* to construct the predictive model:

$$\hat{y}(k+d+1) = g_{ANN}(y(k), \ldots, y(k-n+1), u(k), \ldots u(k-d-m+1); \ W) \tag{2.56}$$

Compared with the recursive *d*-step ahead predictor, the non-recursive predictor is simple for *d*-step ahead prediction as it does not require the recursive procedure. However, a bank of predictors should be used in a predictive horizon if there is any intention to use this type of predictor for long-range prediction. The long-range prediction using non-recursive *i*-step ahead predictor is shown below:

$$\hat{y}(k+i) = g_{ANN}^{i-d}(y(k), \ldots, y(k-n+1), u(k+i-d), \ldots u(k-d-m+1); \ W); \quad d \le i \le P \tag{2.57}$$

In the following, we derive the explicit forms of the aforementioned two neural predictors. The feed-forward neural model, with *q* layers to represent the process can be written as follows:

$$\hat{y}(k+1) = t_f^q\left(W^q \cdot t_f^{q-1}\left(W^{q-1} \cdot \ldots \cdot t_f^1\left(W^1 \cdot I(k)\right)\right)\right) \tag{2.58}$$

where W^j $(j = 1, 2, ..., q)$ is the weight matrix of the network connections between layer $j - 1$ and j; $t_f^j(\cdot)$ is the transfer function of the neurons in layer j, and $I(k) = \{y(k),...,y(k - n + 1, u(k - d),..., u(k - d - m + 1)\}$. We can derive the neural network-based long-range predictor, which is expressed as follows:

$$\hat{y}(k + i|k) = t_f^q \left(W^q \cdot t_f^{q-1} \left(W^{q-1} \cdot \ldots \cdot t_f^1 \left(W^1 \cdot I(k + i - 1) \right) \right) \right); \quad i = 1, \ldots, P$$

(2.59)

where

$I(k + i - 1) = \{\hat{y}(k + i - 1), \ldots, \hat{y}(k + i - n), u(k + i - d), \ldots, u(k + i - d - m)\}$.
If $i - l \leq 0$ $(l = 1,..., n)$, $\hat{y}(k + i - l) = y(k + i - l)$, which is equivalent to the formulation of the argument of the function in Eq. (2.55).

It can be observed that Eq. (2.59) depends on the previous prediction, so that it is a so-called recursive predictor.

For non-recursive neural predictor, the explicit equation can be presented as:

$$\hat{y}(k + i) = t_f^q \left(W_i^q \cdot t_f^{q-1} \left(W_i^{q-1} \cdot \ldots \cdot t_f^1 \left(W_i^1 \cdot I(k + i) \right) \right) \right); \quad i = 1, \ldots, P \quad (2.60)$$

where $I(k + i) = \{y(k), \ldots, y(k + 1 - n), u(k + i - d), \ldots, u(k + 1 - d - m)\}$. It should be noted that, in contrast to the recursive neural predictor, for the different prediction steps within the prediction horizon, the values of the weights of the non-recursive neural predictors are different.

Using the model predictions obtained with one of the procedures described above, the control movement is determined by solving the following optimization problem [128]:

$$\min_{u(k)...u(k+M-1)} \left\{ \sum_{i=1}^P \| Q_i(r(k + i) - \hat{y}(k + i)) \|^2 + \sum_{i=1}^M \| R_i \Delta u(k + i - 1) \|^2 \right\} \quad (2.61)$$

This leads to a non-convex nonlinear optimization problem, for which the global solution is difficult to find; thus, special optimization algorithms should be used. The schematic representation of the DANMPC is presented in Figure 2.12.

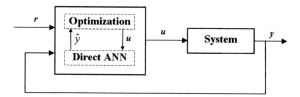

Figure 2.12 Direct artificial neuronal network (ANN) model-based NMPC. The controller is an iterative optimization scheme.

2.3.5.5 Stable DANMPC Control Law

The performance index of the optimization problem from Eq. (2.61) can be written in matrix form as follows:

$$J(k) = E(k)^T \cdot Q \cdot E(k) + \Delta u(k)^T \cdot R \cdot \Delta u(k) \tag{2.62}$$

where $E = r - \hat{y}$ is the model prediction error, k is the discrete time and Q and R respectively are weighting matrices.

A stable control law can be developed to compute the control variable, $u(k)$, so that the performance index is minimized asymptotically. Using the index, $J(k)$ in Eq. (2.62) as a chosen Lyapunov function, clearly $J(k) > 0$ for $E(k) \neq 0$, $\forall k$. The derivative of $J(k)$ w.r.t. k is:

$$\frac{\partial J}{\partial k} = \left(-E^T \cdot Q \cdot \frac{\partial \hat{y}}{\partial u} + \Delta u^T \cdot R \right) \cdot \frac{\partial u}{\partial k} \tag{2.63}$$

It follows that if $(\partial u / \partial k)$ is given as

$$\dot{u} = \frac{\left(\Delta u^T \cdot R - E^T \cdot Q \cdot \frac{\partial \hat{y}}{\partial u} \right)^T \cdot E^T \cdot Q \cdot E}{\left\| \Delta u^T \cdot R - E^T \cdot Q \cdot \frac{\partial \hat{y}}{\partial u} \right\|^2} \tag{2.64}$$

then $\dot{J} = -E^T \cdot Q \cdot E < 0$. This proves that if control increases along the direction given by Eq. (2.64), then index J will converge asymptotically to its minimal value. The calculation of $\dot{u}(t)$ is straightforward, except for $(\partial \hat{y} / \partial u)$, which is according to the structure of the ANN and can have different form.

We illustrate below how to compute this term for a radial basis function network, with the thin-plane-spline non-linear function used as transfer function in the neurons. The ANN model predictions can be obtained by the following equation:

$$\hat{y} = W \cdot \phi \tag{2.65}$$

where $W \in \mathbb{R}^{n_h}$ is the weight matrix with the element w_{rj} connecting the jth hidden layer node to the rth output, n_h the number of hidden layer nodes, and $\phi \in \mathbb{R}^{n_h}$ is the hidden layer output vector with its element given by:

$$\varphi_j = z_j^2 \cdot \ln(z_j), \quad j = 1, \dots, n_h \tag{2.66}$$

with

$$z_j = \| x - C_j \|, \quad j = 1, \dots, n_h \tag{2.67}$$

Here, $x \in \mathbb{R}^q$ is the input vector of the network involving the process input $u \in \mathbb{R}^m$ as m elements, $C_j \in \mathbb{R}^q$, $j = 1, \dots, n_h$, is the jth center vector associated to the jth hidden layer node, and $\|\cdot\|$ is the Euclidean norm.

For this network structure one can obtain:

$$\frac{\partial \hat{y}(t+1)}{\partial u(t)} = \frac{\partial \hat{y}(t+1)}{\partial \phi(t)} \cdot \frac{d\phi(t)}{dz(t)} \cdot \frac{\partial z(t)}{\partial u(t)} \tag{2.68}$$

where

$$\frac{\partial \hat{y}(t+1)}{\partial \phi(t)} = W \tag{2.69}$$

$$\frac{d\phi}{dz} = \begin{bmatrix} z_1 & & \\ & \ddots & \\ & & z_{n_h} \end{bmatrix} \begin{bmatrix} 1 + 2\ln(z_1) & & \\ & \ddots & \\ & & 1 + 2\ln(z_{n_h}) \end{bmatrix} \tag{2.70}$$

$$\frac{\partial z}{\partial u} = \begin{bmatrix} \frac{1}{z_1} & & \\ & \ddots & \\ & & \frac{1}{z_{n_h}} \end{bmatrix} \begin{bmatrix} u_1 - c_{11} & \cdots & u_m - c_{m1} \\ \vdots & \ddots & \vdots \\ u_1 - c_{1n_h} & \cdots & u_m - c_{mn_h} \end{bmatrix} \tag{2.71}$$

with $C_j = [c_{1j}, ..., c_{mj}]^T$ is the jth center. Thus:

$$\frac{\partial \hat{y}}{\partial u} = W \cdot \begin{bmatrix} 1 + 2\ln(z_1) & & \\ & \ddots & \\ & & 1 + 2\ln(z_{n_h}) \end{bmatrix} \begin{bmatrix} u - C_1 & \cdots & u - C_{n_h} \end{bmatrix}^T \tag{2.72}$$

Computation of u(t) according to Eq. (2.64) can be realized in discrete form:

$$u(t+1) = u(t) + \Lambda \cdot t \cdot \dot{u} \tag{2.73}$$

where t is the sampling interval and $\Lambda \in \mathbb{R}^{m \times m}$ is a diagonal matrix with its element $\lambda_{i,i}$ being the learning rate of the ith control, u_i. In order to achieve a fast convergence, an adaptive learning rate should be used to obtain a stable control low.

2.3.5.6 Inverse ANN Model-Based NMPC

Conceptually, the most fundamental neural network-based controllers are probably those using the *inverse* of the process as the controller. The simplest concept is termed *direct inverse control*.

When solving Eq. (2.51) with respect to $u(k)$, one obtains the inverse dynamics of the process:

$$u(k) = f^{-1}(y(k+1), y(k), \dots, y(k-n+1), u(k-1), \dots u(k-m+1)) \tag{2.74}$$

As the function f is unknown, f^{-1} cannot be computed, and thus Eq. (2.74) cannot be evaluated. Nonetheless, an ANN can be trained to approximate the function f^{-1}:

$$\hat{u}(k) = f_{ANN}^{-1}(y(k+1), y(k), \ldots, y(k-n+1), u(k-1), \ldots u(k-m+1); \; W_{inv})$$
(2.75)

where $y(k+1)$, $y(k)$, $\ldots$, $y(k-n+1)$, $u(k-1)$, $\ldots$, $u(k-m+1)$ and $\hat{u}(k)$ are inputs and outputs of the ANN, respectively. By considering a control problem for which the output $y(k)$ tracks a series of setpoints $r(k)$, $k = 0, 1, \ldots$, the inverse model is subsequently applied as controller for the process by inserting the reference $r(k+1)$ as the desired output, instead of the output $y(k+1)$; thus, the off-line optimization problem is eliminated:

$$\hat{u}(k) = f_{ANN}^{-1}(r(k+1), y(k), \ldots, y(k-n+1), u(k-1), \ldots u(k-m+1); \; W_{inv})$$
(2.76)

Usually, in the case of ANN-based control algorithms (even if the exact dynamic model is constructed), an on-line learning must be carried out because the process dynamics may change due to disturbances [129].

The inverse model is subsequently applied as the controller for the process by inserting the desired output, the reference $r(t+1)$, instead of the output $y(t+1)$. Several references are available which use this idea [99,127,130]. A schematic representation of the direct inverse control is presented in Figure 2.13.

Before considering the actual control system, an inverse model must be trained. There are two strategies for obtaining the inverse model, namely *generalized training* and *specialized training* [130]:

- In *generalized training*, a network is trained off-line to minimize the following criterion (W specifies the weights in the network):

$$J(W) = \sum_{k=1}^{N} (u(k) - \hat{u}(k))^2$$
(2.77)

An experiment is performed and a set of corresponding inputs and outputs are stored. Subsequently, the Levenberg–Marquardt training method [131] is invoked.

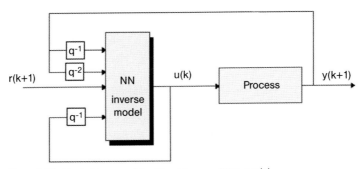

Figure 2.13 Block diagram of the Direct Inverse ANN model-based NMPC.

- *Specialized training* is an on-line procedure related to *model-reference adaptive control*. The idea is to minimize the criterion:

$$J(W) = \sum_{k=1}^{N} (y(k) - y_m(k))^2 \qquad (2.78)$$

where

$$y_m(k) = \frac{q^{-1} \cdot B_m(q)}{A_m(q)} r(k) \qquad (2.79)$$

The inverse model is obtained if $A_m = B_m = 1$, but often a low-pass filtered version is preferred. In this case, the result will be some type of "detuned" (or "smoothed") inverse model.

Specialized training is often said to be goal-directed because, as opposed to generalized training, it attempts to train the network so that the output of the process follows the reference closely. For this reason, specialized training is particularly well suited for optimizing the controller for a prescribed reference trajectory. This is a relevant feature in many robotics applications.

Specialized training must be performed on-line, and thus it is much more difficult to carry out in practice than generalized training. Before the actual training of the inverse model is initiated, a "forward" model of the process must be trained as this is required by the scheme. This can be created from a data set collected in an experiment performed in advance.

Unlike generalized training, the controller design is model-based when the specialized training scheme is applied, as a model of the process is required. Details on the principle can be found in [127]. For minimizing the optimization criterion, different variations of the algorithm have been implemented, for example, recursive back-propagation algorithm or different variations of the recursive Gauss–Newton algorithm.

Specialized training is more complex to implement than generalized training and requires more design parameters. The principle of specialized training is depicted in Figure 2.14.

Although the inverse ANN model-based NMPC are the most alluring ANN-based control techniques, they have a major drawback. Discrete-time neural networks mappings derived from time series can give rise to *multiple* trajectories when followed backwards in time. Consequently, the invertibility of neural networks [132] must be taken into consideration when a controller from this category is designed.

2.3.5.7 ANN Model-Based NMPC with Feedback Linearization

Feedback linearization is a common method for controlling certain classes of nonlinear processes [133]. In order to develop a discrete input-output linearizing controller that is based on a neural network model of the process [134,135], a neural network model with the following structure must be trained:

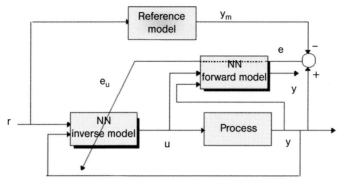

Figure 2.14 The principle of specialized training (model-reference adaptive control).

$$\hat{y}(k) = f(y(k-1), \ldots, y(k-n), u(k-2), \ldots, u(k-m)) + \\ g(y(k-1), \ldots, y(k-n), u(k-2), \ldots, u(k-m)) \cdot u(k-1)$$ (2.80)

where f and g are two separate networks. A feedback linearizing controller is obtained by calculating the controls according to:

$$u(k) = \frac{w(k) - f(y(k), \ldots, y(k-n), u(k-1), \ldots, u(k-m+1))}{g(y(k), \ldots, y(k-n), u(k-1), \ldots, u(k-m+1))}$$ (2.81)

Selecting the virtual control, $w(k)$, as an appropriate linear combination of past outputs plus the reference enables an arbitrary assignment of the closed-loop poles. As for the model-reference controller, feedback linearization is thus a nonlinear counterpart to pole placement with all zeros canceled [136]. The principle of this controller is depicted in Figure 2.15.

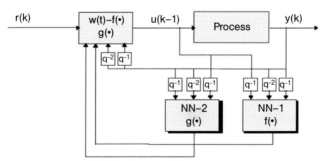

Figure 2.15 Discrete feedback linearization-based NMPC with ANN model.

2.3.5.8 ANN Model-Based NMPC with On-Line Linearization

According to this technique, the ANN model is used to obtain at each sampling instance the linearized process model. The linear model is used then to compute the control low. This approach reduces the complexity of the optimization problem of the NMPC, transforming it to one equivalent with the optimization from the linear MPC.

The ANN model of the process having a form equivalent to Eq. (2.52) is derived:

$$y(t) = f(y(t), \ldots, y(t-n), u(t-d), \ldots u(t-d-m)) \tag{2.82}$$

The *state* $\varphi(t)$ is introduced as a vector composed of the arguments of the function *f*:

$$\varphi(t) = [y(t), \ldots, y(t-n), u(t-d), \ldots u(t-d-m)]^T \tag{2.83}$$

At time $t = \tau$, one can linearize *f* around the current state $\varphi(\tau)$ to obtain the approximate linear model:

$$\tilde{y}(t) = -a_1\tilde{y}(t-1) - \ldots - a_n\tilde{y}(t-n) + b_0\tilde{u}(t-d) + \ldots + b_m\tilde{u}(t-d-m) \tag{2.84}$$

where

$$a_i = -\left.\frac{\partial f(\varphi(t))}{\partial y(t-i)}\right|_{\varphi(t)=\varphi(\tau)} \tag{2.85}$$

$$b_i = \left.\frac{\partial f(\varphi(t))}{\partial u(t-d-i)}\right|_{\varphi(t)=\varphi(\tau)} \tag{2.86}$$

and

$$\tilde{y}(t-i) = y(t-i) - y(\tau-i) \tag{2.87}$$

$$\tilde{u}(t-i) = u(t-i) - u(\tau-i) \tag{2.88}$$

Separating the portion of the expression containing components of the current state vector, the approximate model may alternatively be written as:

$$y(t) = (1 - A(q^{-1})) \cdot y(t) + q^{-d} \cdot B(q^{-1}) \cdot u(t) + \xi(\tau) \tag{2.89}$$

where the *bias* term, $\xi(\tau)$, is determined by:

$$\xi(\tau) = y(\tau) + a_1y(\tau-1) + \ldots + a_ny(\tau-n) - b_0u(\tau-d) - \ldots - b_mu(\tau-d-m) \tag{2.90}$$

and

$$A(q^{-1}) = 1 + a_1q^{-1} + \ldots + a_nq^{-n} \tag{2.91}$$

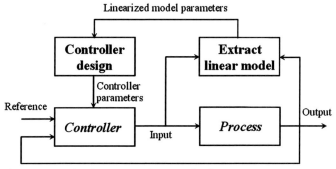

Figure 2.16 On-line linearization for controller design.

$$B(q^{-1}) = b_0 + b_1 q^{-1} + \ldots + b_m q^{-m} \tag{2.92}$$

The approximate model may thus be interpreted as a linear model affected by an additional disturbance, $\xi(\tau)$, depending on the operating point. The idea of applying this principle to the design of the control system is illustrated in Figure 2.16.

2.3.6
NMPC Software for Simulation and Practical Implementation

2.3.6.1 Computational Issues
The prototypical NMPC formulation described in Section 2.4.3 requires that a nonlinear programming problem be solved on-line at each time step in order to determine the manipulated inputs. In general, the optimization problem is non-convex because the model equations are nonlinear. Consequently, the major practical challenge associated with NMPC is on-line solution of the nonlinear program (NLP). Efficient and reliable NLP solution techniques are required to make NMPC a viable control technique. In addition, it may be necessary to derive alternative formulations of the NMPC problem with improved computational properties [137–140]. Some general characteristics of the NLP problem are discussed below, and the most widely studied solution algorithms are presented.

2.3.6.1.1 Nonlinear Programming Problem
The prototypical NMPC formulation is based on a discrete-time state-space model of the nonlinear process. Such a model can be derived by performing state-space implementation on a discrete-time input-output model obtained via nonlinear system identification. In many applications of practical interest an inherently continuous-time nonlinear model derived via fundamental modeling is available for NMPC design. In this case, a discrete-time nonlinear model can be obtained by explicitly discretizing the fundamental model. However, discretization usually is performed implicitly as part of the NLP solution using a numerical technique such as orthogonal collocation [141]. Consequently, continuous-time models should be considered when discussing computational issues [142–145].

For a continuous-time model, the NMPC problem can be represented by Eqs. (2.13) to (2.22). The constraint (2.14) represents the continuous-time model equations over the prediction horizon, while Eq. (2.20) enforces the requirement that all inputs beyond the control horizon are held constant. The constraint (2.16) represents the algebraic equations in the model, while Eqs. (2.20) to (2.22) correspond to the constraints on the input variables, state variables and output variables, respectively. As discussed below, NMPC solution techniques differ primarily according to the method used to handle the model Eqs. (2.14) and (2.16) [76,146–148].

The development of efficient and reliable solution methods for the NLP problem is a challenging problem. The most obvious difficulty is that the optimization problem is nonlinear and has a potentially large number of decision variables. The NLP solution can be computationally too intensive for on-line implementation using conventional process control computers. An equally important problem is that the model constraints (2.14) and (2.16) generally lead to a nonconvex optimization problem [94]. As a result, standard NLP techniques such as successive quadratic programming (SQP) cannot be expected to find the global minimum. Furthermore, there is no theoretical guarantee that any feasible solution can be determined in the presence of nonconvex constraints [149–152].

2.3.6.1.2 Successive Linearization of Model Equations

The simplest way to deal with the model equations (2.14) and (2.16) is to perform *Jacobian linearization* about a nominal operating point and discretize the resulting linear model. This yields linear model predictive control if the objective function is quadratic. Local linearization allows the optimization problem to be solved with simple quadratic programming (QP) techniques, but it provides no compensation for process nonlinearities. A straightforward extension of this idea is to use the current operating point to linearize the model before each execution of the NMPC controller [153]. The primary advantage of successive model linearization is that the NMPC problem is reduced to a LMPC problem at each time step. However, this approach only provides indirect compensation for process nonlinearities.

NMPC techniques based on successive model linearization have been proposed by a number of investigators. Typically, the linearized model is used to predict future process behavior, while the original nonlinear model is used to compute the effect of past input moves [154]. The accuracy of the linear model can be improved by linearizing the model equations several times over the sampling period [155], or by linearizing the model along the computed system trajectory [156]. In the event that the current operating point cannot be determined directly from available process measurements, it becomes necessary to perform the linearization using an estimate of the state variables [157,158]. A related approach is to perform on-line updating of the linear model using the difference between the linear and nonlinear model responses.

2.3.6.1.3 Sequential Model Solution and Optimization

Improved closed-loop performance can be expected if the nonlinear model is used directly in the NMPC calculations. However, standard NLP codes are not designed to handle ODE constraints. This limitation can be overcome using a two-stage solution procedure in which a standard NLP solver is used to compute the manipulated inputs and an ODE solver is used to integrate the nonlinear model equations. This is known as sequential solution because the optimization and integration problems are solved iteratively until the desired accuracy is obtained [159]. First, according to the optimization algorithm, a sequence of control movements is considered; with this sequence, the system of differential equations is numerically integrated to obtain the trajectory of the controlled variables. Then, the scope function is computed. Function of its value, the method of optimization produces a new sequence of control movements and the algorithm is repeated until the optimal sequence is obtained. Only the first value of the sequence is applied to the process. A schematic representation of this approach is shown in Figure 2.17.

As compared to the simultaneous solution method discussed below, an important advantage of the sequential approach is that the manipulated inputs are the only decision variables. Disadvantages of the sequential approach include difficulty in incorporating state/output constraints and poor reliability for large problems.

Several investigators have proposed NMPC solution techniques based on sequential solution of the NLP problem and the model equations. Gradients of the objective function are obtained via simultaneous integration of the model and sensitivity differential equations [160]. The model solution phase can be simplified by discretizing the differential equations (2.14). While other methods are available, the most popular discretization technique is orthogonal collocation on finite elements. This procedure yields nonlinear algebraic model equations of the form: $WX = \Gamma(X, U)$, where X is a matrix of state values at the collocation points, U is a vector of inputs which change over the finite elements, W is a matrix of collocation weights, and Γ is a matrix of nonlinear functions derived from the model function f. Further details relating to this numerical procedure are presented in [161].

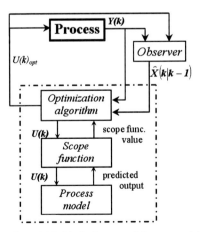

Figure 2.17 Block diagram of the sequential NMPC.

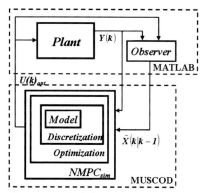

Figure 2.18 Block diagram of the simultaneous NMPC.

2.3.6.1.4 Simultaneous Model Solution and Optimization

An alternative to the sequential solution approach is to solve the optimization problem and the model equations simultaneously. The simultaneous solution method requires the model equations to be discretized since ODEs cannot be handled by standard NLP solvers.

In this approach, the differential equations are discretized and, together with the algebraic ones, they represent the constraints of the optimization function [162]. The optimization problem is solved repeatedly at each sampling point, using a nonlinear programming technique. Figure 2.18 represents the block diagram of this approach. The decision variables are the inputs on each finite element and the state variables at each collocation point. Therefore, the number of decision variables increases as: (i) the sampling period is decreased and/or the prediction horizon is increased (both increase the number of finite elements); or (ii) the number of collocation points on each finite element is increased. The simultaneous approach is best suited for large NLP problems with state/output constraints [163–165].

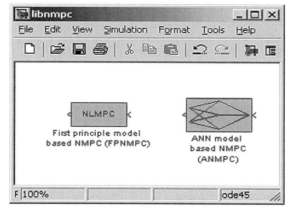

Figure 2.19 SIMULINK library with the block of the nonlinear predictive controller.

2.3.6.2 NMPC Software for Simulation

Although, MATLAB comes with well-developed toolboxes for control system simulations (MPC-T) until now a general-purpose NMPC toolbox has not been developed. The development of a general MATLAB/SIMULINK-based NMPC toolbox, with a different NMPC algorithm, is presented later.

The new NMPC SIMULINK library (Fig. 2.19) contains two main, very general functions:

- *First principle model-based NMPC* (FPNPMC). In this controller, the aforementioned sequential approach was implemented in a MATLAB S-function. Here, each task is performed by a MATLAB function, which has a very general form. It can also be used to simulate NMPC with guaranteed stability (QIH-NMPC). For the optimization, three algorithms were developed. Besides the MATLAB im-

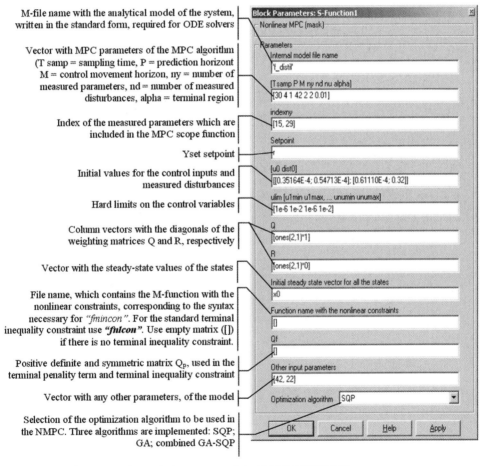

Figure 2.20 The masked parameters of the FPNMPC SIMULINK block.

plementation of the SQP algorithm (the *fmincon* function from the optimization toolbox), two genetic algorithm (GA) -based optimization techniques were developed and implemented. For this, a GA-based optimization toolbox had to be developed (if required, contact the authors for the software). The masked parameters of the FPNMPC block are presented in Figure 2.20.

• *ANN model-based NMPC (ANMPC)*. In this controller the direct and inverse ANN model-based control algorithms are implemented. The new adaptive approach of these controllers in the structure described in distillation control section is also implemented. For optimization, the aforementioned three algorithms were implemented. The ANNMPC block permits the selection of the on-line training algorithm for the adaptive control approaches. For this, besides the Levenberg–Marquardt (LM), Bayesian regularization (BR) and back-propagation (BP), a new combined GA-LM algorithm [166] is also implemented. The interface of the ANMPC block is presented in Figure 2.21.

In addition to the controllers from the NMPC toolbox presented above, a computational efficient FPNMPC was also developed in collaboration with the research

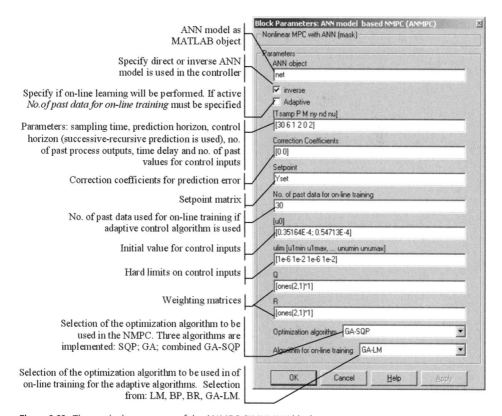

Figure 2.21 The masked parameters of the ANMPC SIMULINK block.

teams from the University of Heidelberg and University of Stuttgart, Germany. This controller uses a special simultaneous approach to solve the control problem. It is based on a tailored version of MUSCOD-II for NMPC [167,168]. MUSCOD-II is an optimization tool that allows the rapid implementation and efficient solution of optimal control problems for ODEs and DAEs using the partially reduced SQP technique. It is based on the multiple shooting method, which is considerably more stable and efficient than the single shooting approach, and allows an effective treatment of both control and state constraints. This controller allows the simulation of FPNMPC with guaranteed stability (both the QIH and zero terminal constraint approaches). An extended Kalman filter (EKF) for the estimation of the unmeasured process states is also introduced in the software package. In our approach, MUSCOD-II was used only for the on-line control problem, which is the most time-consuming task of the NMPC algorithm, the plant simulator and the observer are implemented in MATLAB, and the data exchange is performed through data files. The implementation used for the simultaneous approach is shown schematically in Figure 2.18.

2.3.6.3 NMPC Software for Practical Implementation

The ANN model-based NMPC algorithm was implemented in a general form using the high-level graphical programming language: LabVIEW (National Instruments, SUA). LabVIEW® (Laboratory Virtual Instrument Engineering Workbench) is a

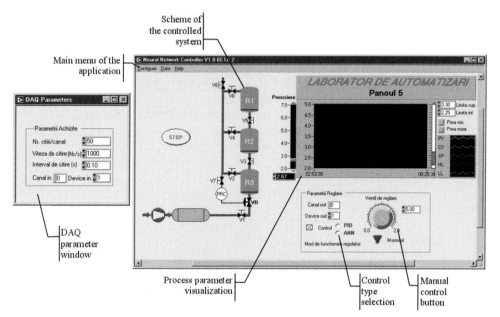

Figure 2.22 The main window of the ANN model-based NMPC software.

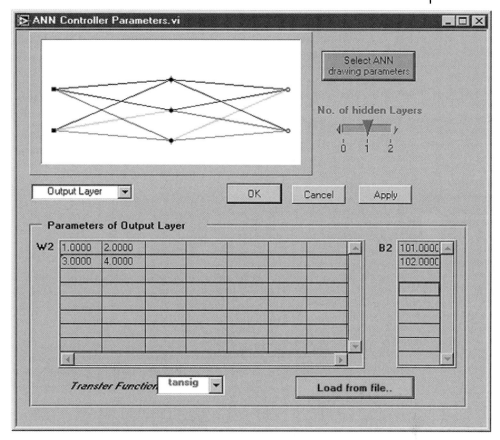

Figure 2.23 ANN model parameter setup window.

powerful instrumentation and analysis software system that departs from the sequential nature of traditional programming languages and creates a graphical programming environment with all the tools needed for data acquisition, analysis, and presentation. With this programming language, programs can be created using a block diagram approach. With its acquisition, analysis, and presentation tools, LabVIEW is functionally complete. Any computation possible in a conventional programming language is possible using the LabVIEW virtual instrument approach [169,170].

This book presents an application to a laboratory chemical plant. The scopes of the experiments are: the study of the dynamic behavior of a cascade of three pressurized vessels, and also the way in which the pressure in the system can be controlled, either manually or by using ANN or PID controller. The main scope of this application is educational, but practical implementation of the ANN model-based controller and the first principle model-based controller for a pilot distillation column is presented in Chapter 3 (Section 3.3). The application provides a detailed description of the equipment and the theoretical background of PID and ANN model-based NMPC control. The LM training algorithm was implemented and can be used for either on-line or off-line identification of the ANN model of the process. The main window of the application is presented in Figure 2.22.

The window for the ANN model parameter selection is presented in Figure 2.23. A particular feature of the developed software consists of the possibility for remote data acquisition (DAQ) and control of the described laboratory process. Although practical experience is a very important part of engineering education, it demands great resources. with the design and construction of state-of-the-art experiments taking time, money, and energy. Consequently, the sharing of experiments both locally and remotely allows unique laboratory equipment to be used at a higher capacity, reduces the experiment cost per student, and offers the student a wider range of experiments.

It is on this basis that the "telelaboratory" paradigm has been implemented. The remote control application is realized in a client/server structure, and consequently the DAQ application, as described above, also acts as a DAQ server. This allows the connection of authenticated clients, and the establishment of a communication with each of the client applications via the TCP/IP protocol. In order to achieve the feeling of "being there" between the DAQ server and client application, an on-line image and sound transfer have been implemented. The main window of the remote client application is presented in Figure 2.24, while a more detailed description can be found in Refs. [171–173].

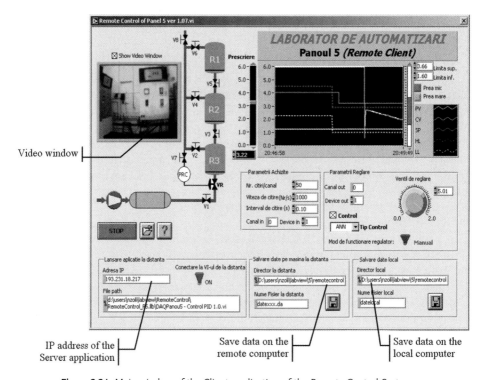

Figure 2.24 Main window of the Client application of the Remote Control System.

2.4
MPC General Tuning Guidelines

The most significant tuning parameters that must be selected for the MPC controllers are *model horizon n, prediction horizon p, control horizon m, sampling interval T, penalty weight matrices* Γ_l^y *and* Γ_l^u, *and a filter.*

The choice of these parameters has a profound effect on the nominal stability, robustness, and controller performance of the MPC algorithms. In applications, the most important criteria that must be satisfied by the controllers are stability and robustness. Thus, it would be convenient to specify a range of control parameters which provides stability and robustness, and then to select from this interval the values for the parameters, which provide the best control performance in accordance with certain control objectives. For linear systems, algorithms are developed to obtain those sufficient conditions which guarantee nominal stability and robustness [174–176]. However, for nonlinear systems the known sufficient conditions [177] are usually much too strong to be met for practical implementation, and one must resort to a set of heuristics based on the extrapolation of linear systems, simulations and experiments.

Hence, the following effects of the tuning parameters on control performance were observed, based on the authors' experience and using examples reported in the literature [178–182].

2.4.1
Model Horizon (*n*)

The model horizon should be selected such that $n \cdot T$ exceeds 95 % of the open-loop settling time of the process. Typical values of *n* vary between 20 and 70. If *n* is too small, truncation problems arise in calculating the predictions, and this can lead to an undesired impulse in the response of the controlled system at time $t = n \cdot T$ when the model error first becomes relevant. It is recommended that a value of *n* such as the last considered value of the impulse response is of the order of magnitude comparable to the measurement error for the output variable.

2.4.2
Prediction Horizon (*p*)

This is generally determined in correlation to the sampling time, *T*. The simple consideration to be taken into account is that the product of the prediction horizon and sampling time, $p \cdot T$, should cover the time necessary for the closed-loop system to achieve steady state (exceeding three to four times the dominant time constant for a first-order process). Typical values for *p* are between 20 and 30. The dimension of the prediction horizon has an important effect on the dimension of the involved MPC matrix, leading to a higher computational effort or even ill-conditioned problems as the dimension increases. In order to overcome this shortcoming, especially for the case of multivariable systems, the prediction horizon may be reduced. Too-short values of p may lead to a more aggressive control action and, in the extreme case, to instability. Usually, a longer prediction horizon leads to a less aggressive control action and slower response. There is a critical minimum horizon length to achieve a stable closed-loop system; this is suggested at the setting $p = n + m$.

2.4.3
Control Horizon (*m*)

When the number of control moves is increased, a more aggressive control action can be observed. This is due to the increased degree of freedom in calculating the control moves that perform a better control at the expense of larger control moves. In this case, the system response is faster, more sensitive to disturbances, and less robust. Additionally, increasing the control horizon leads to increasing dimension of matrices involved in MPC calculations and more complicated optimization problems arise, especially for the nonlinear case. Preserving m as relatively small, with typical values between 1 and 4, conserves good control performance with reduced computational effort and robustness. It is recommended that m be about one-fourth to one-third of the prediction horizon, or alternatively, that $m \cdot T$ should be equal to the time needed for the open-loop response to reach 60% of its steady-state value. For $m = p$, a minimal prototype controller is obtained. Usually, there is an upper band on the control horizon established by the controller performance and the computational complexity of the problem.

2.4.4
Sampling Time (*T*)

In order to ensure good closed-loop performance, the sampling time should be small enough to capture adequately the dynamics of the process, and at the same time large enough to provide feasibility for real-time implementation (the computational time necessary to solve one open-loop control problem should be smaller than the sampling interval). A simple rule of thumb is to choose $T = 0.1(\tau_m + T_d)$, where τ_m and T_d are the dead time and the dominant time constant of the closed loop system, respectively. Zafiriou and Morari proposed some criteria to select the sampling interval for stable linear SISO systems [183]. For unstable systems, robustness depends on the sampling time, and there is a direct relationship between the model error and the maximum allowable size of the sampling time.

2.4.5
Weight Matrices (Γ_j^y and Γ_j^u)

Weighting of the controlled variables is specified in the usually diagonal positive definite matrix, Γ_j^y. The value of the diagonal elements in the Γ_j^y matrix is a measure of the importance of the control effort assigned for each of the controlled variables. As the value of one element from Γ_l^y is increased, the deviation from setpoint of the corresponding controlled variable is decreased.

Adding in the performance index (a term for penalizing the movement of the manipulated variables) reduces the excessive manipulated variable move. This is performed by use of the usually diagonal positive definite matrix Γ_l^u for weighting the manipulated variable move. Increasing the value of one element from Γ_l^u decreases the corresponding manipulated variable change with effect on degradation of the control performance but increasing robustness.

Some attempts to obtain these parameters have been presented in the literature [184].

It is important to note that the relative magnitudes of Γ_l^y and Γ_l^u have to be considered as the magnitudes for the controlled and manipulated variables may be of different order of magnitude.

The weight matrices Γ_l^y and Γ_l^u may be considered nondiagonal, as interactions between controlled variables are important and time-varying as control performance has to be dynamically changed. Unfortunately, there are no reliable tuning guides for these approaches due to the hidden influences between the tuning parameters and the control performance.

2.4.6
Feedback Filter

The use of a well-tuned filter on the feedback signal provides good disturbance rejection and fast system response. However, the choice and effects of the filter depend heavily on the certain system [185]. An example of the effects of the Extended Kalman Filter (EKF) used in the NMPC of a high-purity distillation column is presented in Chapter 3 (Section 3.3).

2.4.7
Dynamic Sensitivity Used for MPC Tuning

Usually, tuning difficulties are related to the MIMO characteristics of the control problem, and to the poor prediction of the effect of tuning parameters on control performance criteria. These difficulties become more important when nonlinear behavior of the model is present. Due to this, the model predictive controller tuning may become an iterative character and control performance may be enhanced, to a large extent, by repeated simulations (i.e., involving heuristic approaches).

For evaluating the closed-loop influence of the generic tuning parameter p on (controlled) variable y, the sensitivity functions may be determined according to equation:

$$S_p^y(t) = \left[S_{p_j}^{y_i} \right] = \left[\frac{\partial y_i}{\partial p_j} \frac{p_j}{y_i} \right] \tag{2.93}$$

where S_p^y is the matrix of the scaled sensitivity of variable y_i with respect to parameter p_j, calculated for step initial variations.

The dynamic sensitivity of both the controlled variables and of the square error (between the controlled variables and the setpoint, i.e., the performance index) with respect to the tuning parameters may be determined. The considered tuning parameters may be: the prediction horizon p; the number of manipulated variable moves (input horizon) m; and the elements of the diagonal input and output weighting matrices Γ_l^u, Γ_l^y of the MPC optimization objective function. Useful information for parameter tuning can be obtained by analyzing the value of the sensitivity of the square error. Usually, small values of these sensitivity functions with respect to different tuning parameters denote appropriate tuning. It is important to note that the scaled sensitivity functions offer a quantitative measure for evaluating the influence of tuning parameters on the quality of control (square error index). As a consequence, a direct indication of the tuning parameter to be changed, as well as its best value, may be obtained [181,182].

References

1 Newton, G. C., Gould, L. A., Kaiser, J. F., *Analytical Design of Feedback Controls*. Wiley, New York, **1957**.

2 Smith, O. J. M., Closer control of loops with dead time. *Chem. Engng. Prog.*, **1957**, *53*(5), 217–219.

3 Frank, P. M., *Entwurf von Regelkreisen mit vorgeschriebenen Verhalten*. G. Braun, Karlsruhe, **1974**.

4 Brosilow, C. B., The structure and design of Smith Predictors from the viewpoint of inferential control. *Proc. Joint Aut. Control Conf.*, Denver, Colorado, **1979**.

5 Garcia, C. E., Morari M., Internal model control-1. A unifying review and some new results, *Ind. Eng. Chem. Process Des. & Dev.*, **1982**, *21*, 308–323.

6 Garcia, C. E., Morari M., Internal model control-3. Multivariable control law computation and tuning guidelines. *Ind. Engng. Chem. Process Des. Dev.*, **1985**, *24*, 484–494.

7 Morari, M., Zafiriou, E., *Robust Process Control*, Prentice-Hall, Englewood Cliffs, New Jersey, **1989**.

8 Merlin, T. E., *Process control: designing processes and control systems for dynamic performance*, 2nd edition 2000, McGraw-Hill, publisher Thomas Casson.

9 Brosilow C., Joseph B., *Techniques of Model-based Control*, Prentice Hall PTR, New Jersey, **2002**.

10 Zadeh, L. A., Whalen, B. H., On optimal control and linear programming, *IRE Trans. Aut. Control*, **1962**, *7*(4), 45.

11 Propoi, A. I., Use of LP methods for synthesizing sampled-data automatic systems. *Autom. Remote Control*, **1963**, *24*, 837.

12 Richalet, J. A., Rault, A., Testud, J. T., Papon, J., Model predictive heuristic control: applications to an industrial process, *Automatica*, **1978**, *14*, 413–428.

13 Cutler, C. R., Ramaker, B. L., Dynamic matrix control – a computer control algorithm. AIChE National Meeting, Houston, Texas, **1979**.

14 Prett, D. M., Gillette, R. D., Optimization and constrained multivariable control of a catalytic cracking unit, AIChE National Mtg. Houston, Texas, **1979**.

15 Garcia, C. E., Prett, D. M., Morari, M., Model predictive control: theory and practice – a survey, *Automatica*, **1989**, *25*(3), 335–348.

16 Economou, C. G., Morari, M., Palsson, B., Internal model control-5, extension to nonlinear systems. *Ind. Engng. Chem. Process Des. Dev.*, **1986**, *25*, 403–411.

17 Eaton, J. W., Rawlings, J. B., Model-predictive control of chemical processes, *Chemical Engineering Science*, **1992**, *47*(4), 705–720.

18 Clarke, D. W., Mohtadi, C., Tuffs, P. S., Generalized predictive control – Part I. The basic algorithm, *Automatica*, **1987**, *23*(2), 137–148.

19 Clarke, D. W., Mohtadi, C., Tuffs, P. S., Generalized predictive control – Part II. Extensions and interpretations, *Automatica*, **1987**, *23*, (2), 149–160.

20 Demircioglu, H., Gawthrop, P. J., Continuous-time Generalized Predictive Control (CGPC), *Automatica*, **1991**, *27*(1), 55–74.

21 Morari, M., Lee, J. H., Model predictive control: past, present and future. *Computers and Chemical Engineering*, **1999**, *23*, 667–682.

22 Aufderheide, B., Bequette, B. W., Extension of dynamic matrix control to multiple models, *Computers and Chemical Engineering* , **2003**, *27*, 1079–1096.

23 Bindlish, R., Rawlings, J. B., Target linearization and model predictive control of polymerization processes, *AIChE Journal*, **2003**, *49*(11), 2885–2899.

24 Odloak, D., Extended robust model predictive control, *AIChE Journal*, **2004**, *50*(8), 1824–1836.

25 Guiamba, I. R. F., Mulholland, M., Adaptive Linear Dynamic Matrix Control applied to an integrating process, *Computers and Chemical Engineering*, **2004**, *28*(12), 2621–2633.

26 Eker, S. A., Nikolaou, M., Linear control of nonlinear systems: Interplay between nonlinearity and feedback, *AIChE Journal*, **2002**, *48*(9), 1957–1980.

27 Bacic, M., Cannon, M., Lee, Y. I., Kouvaritakis, B., General interpolation in MPC and its advantages. *IEEE Transactions on Automatic Control*, **2003**, *48*(6), 1092–1096.

28 Saffer II, D. R., Doyle III, F. J., Analysis of linear programming in model predictive control, *Computers and Chemical Engineering*, **2004**, *28*, 2749–2763.

29 Seborg, D. E., Edgar, T. F., Mellichamp, D. A., *Process Dynamics and Control*, John Wiley & Sons, 649–669, **1989**.

30 Muske, K. R., Rawlings, J. B., Model predictive control with linear models, *AIChE Journal*, **1993**, *39*(2), 262–287.

31 Lee, J. H., Morari, M., Garcia, C. E., State-space interpretation of model predictive control, *Automatica*, **1994**, *30*(4), 707–717.

32 Balchen, J. G., Ljungquist, D., Strand, S., State space predictive control, *Chem. Eng. Sci.*, **1992**, *47*(4), 787–807.

33 Rodrigues, M. A., Odloak, D., An infinite horizon model predictive control for stable and integrating processes, *Computers and Chemical Engineering* , **2003**, *27*, 1113–1128.

34 Grieder, P., Borrelli, F., Torrisi, F., Morari, M., Computation of the constrained infinite time linear quadratic regulator, *Automatica*, **2004**, *40*, 701–708.

35 Carrapiço, O. L., Odloak, D., A stable model predictive control for integrating processes, *Computers and Chemical Engineering*, **2005**, *29*(5),1089–1099.

36 Grancharova, A., Johansen, T. A., Kocijan, J., Explicit model predictive control of gas–liquid separation plant via orthogonal search tree partitioning, *Computers and Chemical Engineering*, **2004**, *28*, 2481–2491.

37 Hegrenæs, O., Gravdahl, J. T., Tø|«ndel, P., Spacecraft attitude control using explicit model predictive control, *Automatica*, **2005**, *41*, 2107–2114.

38 Mhaskar, P., El-Farra, N. H., Christofides, P. D., Hybrid predictive control of process systems, *AIChE Journal*, **2004**, *50*(6), 1242–1259.

39 Allgower F., Badgwell, T. A., Qin, J. S., Rawlings, J. B., Wright, S. J., Nonlinear predictive control and moving horizon estimation – An introductory overview. In: Frank P. M. (Ed.), *Advances in Control, Highlights of ECC '99*, Springer, pp. 391–449, **1999**.

40 Henson, A. M., Nonlinear model predictive control: current status and future directions, *Computers and Chemical Engineering*, **1998**, *23*, 187–201.

41 Imsland, L., Findeisen, R., Bullinger, E., Allgower, F., Foss, B. A., A note on stability, robustness and performance of output feedback nonlinear model predictive control, *Journal of Process Control*, **2003**, *13*, 633–644.

42 Jianping, G., Patwardhan, R., Akamatsu, K., Hashimoto, Y., Emoto, G., Sirish L. S., Huang, B., Performance evaluation of two industrial MPC controllers, *Control Engineering Practice*, 2003, *11*, 1371–1387.

43 Santos, L. O., Paulo A. F. N. A., Afonso, J. A. A.M., Castro, N. M. C., Biegler, L. T., On-line implementation of non-linear MPC: an experimental case study, *Control Engineering Practice*, **2001**, *9*, 847–857.

44 Mayne, D. Q., Non-linear model predictive control: challenges and opportunities. In: Allgöwer, F., Zheng, A. (Eds), *Nonlinear Model Predictive Control*, Birkhauser, Basel, pp. 23–44, **2000**.

45 Nikolaou, M., Model predictive controllers: a critical synthesis of theory and industrial needs, *Advances in Chemical Engineering Series*, **2001**.

46 McAvoy, T., Intelligent control applications in the process industries, *Annual Reviews in Control*, **2002**, *26*, 75–86.

47 Qin, S. J., Badgwell, T. A., An overview of nonlinear model predictive control applications. In: Allgower, F., Zheng, A. (Eds), *Nonlinear Predictive Control*, Birkhauser, Basel, **2000**.

48 Poloski, A. P., Kantor, J. C., Application of model predictive control to batch processes, *Computers and Chemical Engineering*, **2003**, *27*, 913–926.

49 Bequette, B. W., Nonlinear control of chemical processes – a review, *Ind. Eng. Chem. Res.*, **1991**, *30*, 1391–1413.

50 Morari, M. and Lee, J. H., Model predictive control: Past, present and future. In: *Proceedings PSE »97-ESCAPE-7 Symposium*, Trondheim, **1997**.

51 Lee, J. H., Cooley, B., Recent advances in model predictive control and other related areas. In: Kantor, J. C., Garcia, C. E., Carnahan, B. (Eds), *Fifth International Conference on Chemical Process Control – CPC V*, American Institute of Chemical Engineers, pp. 201–216, **1996**.

52 Galan, O., Romagnoli J. A., Palazoglu, A., Robust H∞ control of nonlinear plants based on multi-linear models: an

application to a bench-scale pH neutralization reactor, *Chem. Eng. Sci.*, **2000**, *55*, 4435–4450.

53 Nagy, Z., Agachi, S., Dynamic modelling and model predictive control of a fermentation bioreactor, *Proceedings on the International Chemistry Show of the Romanian Society of Chemical Engineers, SiChem »98*, October 20–23, Bucharest, Romania, pp. 350–359, **1998**.

54 De Oliveira, S. L., *Model predictive control for constrained nonlinear systems*, PhD thesis, Swiss Federal Institute of Technology (ETH), Zurich, Switzerland, **1996**.

55 Muske, K. R., Howse, J. W., Hansen, G. A., Cagliostro, D. J., Model-based control of a thermal regenerator. Part 1: dynamic model, *Computers and Chemical Engineering*, **2000**, *24*, 2519–2531.

56 Muske, K. R., Howse, J. W., Hansen, G. A., Cagliostro, D. J., Model-based control of a thermal regenerator. Part 2: control and estimation, *Computers and Chemical Engineering*, **2000**, *24*, 2507–2517.

57 Brengel, D. D., Geider, W. D., Multistep nonlinear predictive controller, *Ind. Eng. Chem. Res.*, **1989**, *20*, 1812–1822.

58 Nagy, Z., Agachi, S., Model predictive control of a PVC batch reactor, *Computers and Chemical Engineering*, **1997**, *21*(6), 571–591.

59 Qin, S. J., Badgewell, T., A survey of industrial model predictive control technology, *Control Engineering Practice*, **2003**, *11*, 733–764.

60 Franke, R., Formulation of dynamic optimization problems using Modelica and their efficient solution. In: *Proceedings of the 2ⁿᵈ International Modelica Conference*, pp. 315–323, **2002**.

61 Franke, R., Rode, M., Krueger, K., Online optimization of drum boiler startup. In: *Proceedings of the 3ʳᵈ International Modelica Conference*, pp. 287–296, **2003**.

62 Ogunnaike, B., Wright, R. A., Industrial applications of nonlinear control. In: Kantor, J. C., Garcia, C. E., Carnahan, B. (Eds), *Fifth International Conference on Chemical Process Control, Volume 93, AIChE Symposium Series No. 316*, pp. 46–59, New York, **1997**.

63 Van Brempt, W., Backx, T., Ludlage, J., Van Overschee, P., De Moor, B., Tousain, R., A high performance model predictive controller: application on a polyethylene gas phase reactor, *Control Engineering Practice*, **2001**, *9*, 829–835.

64 Gopaluni, R. B., Patwardhan, R. S., Shah, S. L., The nature of data pre-filters in MPC relevant identification-open- and closed-loop issues, *Automatica*, **2003**, *39*, 1617–1626.

65 Shouche, M. S., Genceli, H., Nikolaou, M., Effect of on-line optimization techniques on model predictive control and identification (MPCI), *Computers and Chemical Engineering*, **2002**, *26*, 1241–1252.

66 Nagy, Z. K., Braatz, R. D., Robust nonlinear model predictive control of batch processes, *AIChE Journal*, **2003**, *49*(7), 1776–1786.

67 Wang, D., Romagnoli, J. A., Robust model predictive control design using a generalized objective function, *Computers and Chemical Engineering*, **2003**, *27*, 965–982.

68 Sung-Mo, A., Park, M.-J., Rhee, H.-K., Extended Kalman filter-based nonlinear model predictive control for a continuous MMA polymerization reactor, *Ind. Eng. Chem. Res.*, **1999**, *38*, 3942–3949.

69 Biegler, L., Efficient solution of dynamic optimization and NMPC problems. In: Allgower F. and Zheng A. (Eds), *Nonlinear Predictive Control*, Bagel, Birkhauser, **2000**.

70 Biegler, L. T., Solution of dynamic optimization problems by successive quadratic programming and orthogonal collocation, *Computers and Chemical Engineering*, **1984**, *8*, 243.

71 Biegler, L. T., Rawlings, J. B., Optimisation approaches to nonlinear model predictive control, *Proceedings of Conference on Chemical Process Control*, South Padre Island, Texas, pp. 543–571, **1991**.

72 Bock, H. G., Bauer, I., Leineweber, D., Schloder, J., Direct multiple shooting methods for control and optimization in engineering. In: Keil, F., Mackens, W., Vos, H., Werther, J. (Eds), *Scientific Computing in Chemical Engineering II*, Volume 2, Springer, **1999**.

73 Hassapis, G., Implementation of model predictive control using real-time multiprocessing computing, *Microprocessors and Microsystems*, **2003**, *27*, 327–340.

74 De Oliveira, N. M. C., Biegler, L. T., Constraint handling and stability properties of model-predictive control, *AIChE Journal*, **1994**, *40*(2), 1138–1155.

75 Lee J. H., Modeling and identification for nonlinear model predictive control: requirements, current status and future research needs. In: Allgower, F., Zheng, A. (Eds), *Proceedings of International Symposium on Nonlinear Model Predictive Control: Assessment and Future Directions*, Ascona, Switzerland, pp. 91–107, **1998**.

76 De Oliveira, N. M. C., Biegler, L. T., An extension of Newton-type algorithms for nonlinear process control, *Automatica*, **1995**, *31*(2), 281–286.

77 Campo, P. J., Morari, M., inf-Norm formulation of model predictive control problems. *Proceedings American Control Conference*, Seattle, Washington, pp. 339–343, **1986**.

78 Keerthi, S. S., Gilbert, E. G., Optimal infinite-horizon feedback laws for a general class of constrained discrete-time systems: Stability and moving-horizon approximations, *Journal of Optimization Theory and Applications*, **1988**, *57*(2), 265–293.

79 Chen, H., Allgower, F., A quasi-infinite horizon predictive control scheme for constrained nonlinear systems. In: *Proceedings 16th Chinese Control Conference*, Qindao, pp. 309–316, **1996**.

80 Chen, H., Allgower, F., Nonlinear model predictive control schemes with guaranteed stability. In: Berber, C., Kravaris, R. (Eds), *Nonlinear Model Based Process Control*, Kluwer Academic Publishers, Dodrecht, pp. 465–494, **1998**.

81 Findeisen, R., Allgower, F., A procedure to determine a region of attraction for discrete time nonlinear systems, Technical Report # 99.14, Automatic Control Laboratory, Swiss Federal Institute of Technology (ETH), Zürich, Switzerland, April, **1999**.

82 Findeisen, R., Rawlings, J. B., Suboptimal infinite horizon nonlinear model predictive control for discrete time systems, Technical Report # 97.13, Automatic Control Laboratory, Swiss Federal Institute of Technology (ETH), Zürich, Switzerland, presented at the NATO Advanced Study Institute on Nonlinear Model Based Process Control, **1997**.

83 Sistu, P. B., Bequette, B. W., Nonlinear model-predictive control: closed-loop stability analysis, *AIChE Journal*, **1996**, *42*, 3388–3202.

84 De Nicolao G., Magni, L., Scattolini, R., On the robustness of receding horizon control with terminal constraints, *IEEE Trans. Automat. Contr.*, **1996**, *41*(3), 451–453.

85 De Nicolao G., Magni, L., Scattolini, R., Stabilizing receding-horizon control of nonlinear time-varying systems. In: *Proceedings, 4th European Control Conference ECC »97*. Brussels, **1997**.

86 De Nicolao, G., Magni, L., Scattolini, R., Stability and robustness of nonlinear receding horizon control. In: Allgower, F., Zheng, A. (Eds), *Proceedings of International Symposium on Nonlinear Model Predictive Control: Assessment and Future Directions*, Ascona, Switzerland, pp. 77–90, **1998**.

87 Chen, H., Allgower, F., A quasi-infinite horizon nonlinear predictive control scheme for stable systems: Application to a CSTR. In: *Proceedings International Symposium on Advanced Control of Chemical Processes*, ADCHEM, Banff, Canada, pp. 471–476, **1997**.

88 Chen H., Scherer, C. W., Allgower, F., A game theoretic approach to nonlinear robust receding horizon control of constrained systems. In: *Proceedings American Control Conference*, Albuquerque, pp. 3073–3077, **1997**.

89 Chen, H., Scherer, C. W., Allgower, F., A robust model predictive control scheme for constrained linear systems. In: *Fifth IFAC Symposium on Dynamics and Control of Process Systems*, DYCOPS-5, Korfu, pp. 60–65, **1998**.

90 Findeisen, R., Allgower, F., Nonlinear model predictive control for index-one DAE systems. In: Allgower, F., Zheng, A. (Eds), *Nonlinear Predictive Control*, Birkhauser, **2000**.

91 Mayne, D. Q., Michalska, H., Receding horizon control of nonlinear systems, *IEEE Trans. on Automatic Control, 1990, 35*(7), 814–824.

92 Michalska, H., Mayne, D. Q., Robust receding horizon control of constrained nonlinear systems, *IEEE Trans. on Automatic Control, 1993, 38*, 1623–1633.

93 Chisci, L. L., Lombardi, A., Mosca, E., Dual-receding horizon control of constrained discrete time systems, *European Journal of Control*, **1996**, *2*, 278–285.

94 Mayne, D. Q., Nonlinear model predictive control: An assessment. In: Kantor, J. C., Garcia, C. E., Carnahanr, B. (Eds), *Fifth International Conference on Chemical Process Control -CPC V*, American Institute of Chemical Engineers, pp. 217–231, **1996**.

95 Mayne, D. Q., Nonlinear model predictive control: Challenges and opportunities. In: Allgower, F., Zheng, A. (Eds), *Nonlinear Predictive Control*, Birkhauser, Basel, **2000**.

96 Meadows, E. S., Henson, M. A., Eaton, J. W., Rawlings, J. B., Receding horizon control and discontinuous state feedback stabilization, *International Journal of Control*, **1995**, *62*(5), 1217–1229.

97 Meadows, E. S., Rawlings, J. B., Receding horizon control with an infinite horizon. In: *Proceedings American Control Conference*, San Francisco, pp. 2926–2930, **1993**.

98 Chen, H., Allgower, F., A quasi-infinite horizon nonlinear model predictive control scheme with guaranteed stability, *Automatica, 1998, 34*(10), 1205–1218.

99 Hunt, K. J., Sbarbaro, D., Zbikowski, R., Gawthrop, P. J., Neural Networks for Control Systems – A survey, *Automatica*, **1992**, *28*(6), 1083–1112.

100 Aldrich, C., van Deventer, J., Comparison of different artificial neural nets for the detection and location of gross errors in process systems, *Ind. Eng. Chem. Res.*, **1995**, *34*(1), 216–224.

101 Ungar, L. H., Powel, B., Kamens, S. N., Adaptive network for fault diagnosis and process control, *Comp. Chem. Eng.*, **1990**, *14*, 561–572.

102 Di Massimo, C., Montague, G. A., Willis, M. J., Tham, M. T., Morris, A. J., Towards improved penicillin fermentation via artificial neural networks, *Comp. Chem. Eng.*, **1992**, *16*(4), 283–291.

103 Nascimento, C. A. O., Giudici, R., Guardani, R., Neural network based approach for optimization of industrial chemical processes, *Comp. Chem. Eng.*, **2000**, *24*, 2303–2314.

104 Baughman, D. R., Liu, Y. A., An expert network for predictive modeling and optimal design of extractive bioseparations in aqueous two-phase systems, *Ind. Eng. Chem. Res.*, **1994**, *33*(11), 2668–2687.

105 Agachi, S., Kiss, A. A., Nagy, Z., Pop, A., Modeling of catalytic steam reforming process using neural networks, *Studia Universitatis Babes-Bolyai, Ser. Chemia*, **1998**, *43*(1–2), 135–144.

106 Bhat, N. V., McAvoy, T. J., Use of neural nets for dynamic modeling and control of chemical process systems, *Computers and Chemical Engineering*, **1990**, *14*(4/5), 573–585.

107 Cheng, Y., Karjala, T. W., Himmelblau, D. M., Identification of nonlinear dynamic processes with unknown and variable dead time using an internal recurrent neural network, *Ind. Eng. Chem. Res.*, **1995**, *34*(5), 1735–1742.

108 Nagy, Z., Cormos, A.-M., Agachi, S., Stabilirea consumurilor specifice de cocs în procesul de ardere a calcarului cu reţele neuronale artificiale (RNA), *Revista de Chimie*, **2001**, 6, 330–334.

109 Pollard, J. F., Broussard, M. R., Garrison, D. B., San, K. Y., Process Identification Using Neural Networks, *Computers and Chemical Engineering*, **1992**, *16*(4), 253–270.

110 Gao, F., Wang, F., Li, M., A simple nonlinear controller with diagonal recurrent neural network, *Chem. Eng. Sci.*, **2000**, 55, 1283–1288.

111 Hoskins, J. C., Himmelblau, D. M., process control via artificial neural networks and reinforcement learning, *Computers and Chemical Engineering*, **1992**, *16*(4), 241–251.

112 Joseph, B., Hanratty, F. W., Predictive control of quality in a batch manufacturing process using artificial neural network models, *Ind. Eng. Chem. Res.*, **1993**, *32*(9), 1951–1961.

113 Willis, M. J., Montague, G. A., Di Massimo, C., Tham, M. T., Morris, A. J., Artificial neural networks in process estimation and control, *Automatica*, **1992**, *28*(6), 1181–1187.

114 Kavuri, S. N., Venkatasubramanian, V., Combining pattern classification and assumption-based techniques for process fault diagnosis, *Computers and Chemical Engineering*, **1992**, *16*(4), 299–312.

115 Whiteley, J. R., Davis, J. F., Knowledge-based interpretation of sensor patterns, *Computers and Chemical Engineering*, **1992**, *16*(4), 329–346.

116 Rustoiu-Csavdari, A., Copolovici, L., Nagy, Z., Possible use of artificial neural networks in tricomponent calibration, *14th International Congress of Chemical and Process Engineering CHISA »2000*, 27–31 August, Praha, full text on CD, Paper No. 1447 (P7.4), **2000**.

117 Zhang, B. S., Leigh, J. R., Porter, N., Hill, D., Application of statistical and neural network techniques to biochemical data analysis. In: *Proceedings American Control Conference*, Albuquerque, pp. 3267–3268, **1997**.

118 Hernandez, E., Arkun, Y., Study of the control-relevant properties of backpropagation neural network models of nonlinear dynamic systems, *Computers and Chemical Engineering*, **1992**, *16*(4), 227–240.

119 Chen, J., Bruns, D. P., WaveARX neural network development for system identification using a systematic design synthesis, *Ind. Eng. Chem. Res.*, **1995**, *34*(12), 4420–4435.

120 Kramer, M. A., Autoassociative neural networks, *Computers and Chemical Engineering*, **1992**, *16*(4), 313–328.

121 Mavrovouniotis, M. L., Chang, S., Hierarchical neural networks, *Computers and Chemical Engineering*, **1992**, *16*(4), 347–369.

122 Bhat, N. V., McAvoy, T. J., Determining model structure for neural models by network stripping, *Computers and Chemical Engineering*, **1992**, *16*(4), 271–281.

123 Demuth, H., Beale, M., (Eds), *Neural Network Toolbox User's Guide*, The MathWorks, Inc., **1993**.

124 Chen, P. C. Y., Mills, J. K., A methodology for analysis of neural network generalization in control systems. In: *Proceedings American Control Conference*, Albuquerque, pp. 1091–1095, **1997**.

125 Foresee, F. D., Hagan, M. T., Gauss-Newton approximation to Bayesian regularization. In: *Proceedings of the 1997 International Joint Conference on Neural Networks*, pp. 1930–1935, **1997**.

126 Nagy, Z., Agachi, S., Bodizs, L., Adaptive neural network model based nonlinear predictive control of a fluid catalytic cracking unit, *European Symposium on Computer Aided Process Engineering-10, ESCAPE-10*, Florence, Italy. In: Pierucci, S. (Ed.), *European Symposium on Computer Aided Process Engineering-10*, Elsevier Science, pp. 235–240, **2000**.

127 Hunt, K. J., Sbarbaro, D., Neural networks for nonlinear internal model control, *IEE Proceedings-D*, **1991**, *138*(5), 431–438.

128 Long, T. W., Hanzevack, E. L., Midwood, B. R., A neural network based receding horizon optimal (RHO) controller. In: *Proceedings American Control Conference*, Albuquerque, pp. 1994–1995, **1997**.

129 Ge, S. S., Hang, C. C., Zhang, T., Direct adaptive neural network control of nonlinear systems. In: *Proceedings American Control Conference*, Albuquerque, pp. 1568–1572, **1997**.

130 Psaltis, D., Sideris, A., Yamamure, A. A., A multilayered neural network controller, *Control Sys. Mag.*, **1988**, *8*(2), 17–21.

131 Fletcher, R., *Practical Methods of Optimization*, Wiley, **1987**.

132 Rico-Martinez, R., Adomaitis, R. A., Kevrekidis, I. G., Noninvertibility in neural networks, *Computers and Chemical Engineering*, **2000**, *24*, 2417–2433.

133 Rawlings, J. B., Muske, K. R., Nonlinear predictive control of feedback linearisable systems and flight control system design, *IEEE Trans. on Automatic Control*, **1993**, *38*(10), 1512–1516.

134 Ayala Botto, M., van den Boom, T. J. J., Krijgsman, A., Sa da Costa, J. M. G., Predictive control based on neural network models with I/O feedback linearization, *International Journal of Control*, **1999**, *72*, 1538–1554.

135 Ge, S. S., Robust adaptive NN feedback linearization control of nonlinear systems. In: *Proceedings of IEEE International Symposium on Intelligent Control*, Ritz-Carlton, Dearborn, Michigan, USA, pp. 486–491, **1996**.

136 Astrom, K. J., Wittenmark, B., *Adaptive Control*, Addison-Wesley, **1995**.

137 Sistu, P. B., Gopinath, R. S., Bequette, B. W., Computational issues in nonlinear predictive control, *Computers and Chemical Engineering*, **1993**, *17*(4), 361–366.

138 Wright, G. T., Edgar, T. F., Nonlinear model predictive control of a fixed-bed water-gas shift reactor: an experimental study, *Computers and Chemical Engineering*, **1994**, *18*, 83–102.

139 Wright, S. J., Applying new optimization algorithms to model predictive control. In: Kantor, J. C., Garcia, C. E., Carnahan, B. (Eds), *Fifth International Conference on Chemical Process Control-CPC V*, American Institute of Chemical Engineers, pp. 147–155, **1996**.

140 Wright, S. J., *Primal-Dual Interior-Point Method*, SIAM Publications, Philadelphia, **1997**.

141 M. Diehl, *Real-Time Optimization for Large Scale Nonlinear Processes*. PhD Thesis, University of Heidelberg, **2001**.

142 Bellman, R. E., Dreyfus, S. E., *Applied Dynamic Programming*, Princeton University Press, Princeton, New Jersey, **1962**.

143 Bertsekas, D. P., *Nonlinear Programming*, Athena Scientific Press, Belmont, MA, **1995**.

144 Binder, T., Blank, L., Bock, H. G., Bulitsch, R., Dahmen, W., Diehl, M., Kronseder, T., Marquardt, W., Schloeder, J., and von Stryk, O., Introduction to model based optimization of chemical processes on moving horizons. In: Groetschel, M., Krumke, S. O., Rambau, J. (Eds.), *Online Optimization of Large Scale Systems*. Berlin, Springer, **2001**.

145 Biegler, L. T., Rawlings, J. B., Optimisation approaches to nonlinear model predictive control. In: *Proceedings of Conference on Chemical Process Control*, South Padre Island, Texas, pp. 543–571, **1991**.

146 Bock, H. G., Bauer, I., Leineweber, D., Schloder, J., Direct multiple shooting methods for control and optimization in engineering. In: Keil, F., Mackens, W., Vos, H., Werther, J. (Eds), *Scientific Computing in Chemical Engineering II*, Volume 2, Springer, **1999**.

147 Bock, H. G., Diehl, M., Schloder, J. P., Allgower, F., Findeisen, R., Nagy, Z., Real-time optimization and nonlinear model predictive control of processes governed by differential-algebraic equations. In: *Proceedings of the International Symposium on Advanced Control of Chemical Processes, ADCHEM 2000*, Pisa, Italy, pp. 695–703, **2000**.

148 Bock, H. G., Diehl, M., Leineweber, D., Schloder, J., A direct multiple shooting method for real-time optimization of nonlinear DAE processes. In: Allgower, F., Zheng, A. (Eds), *Nonlinear Predictive Control*, Birkhauser, **2000**.

149 Economou, C. G., Morari, M., Newton control laws for nonlinear controller design. In: *Proceedings, IEEE Conference on Decision and Control*, Fort Lauderdale, Florida, pp. 1361–1366, **1985**.

150 Garcia, C. E., Morshedi, A. M., Quadratic programming solution of dynamic matrix control (QDMC). In: *Proceedings American Control Conference*, San Diego, California; also *Chem. Engng Commun.*, **1986**, *46*, 73–87.

151 Kiparissides, C., Georgiou, A., Finite-element solution of nonlinear optimal control problems with a quadratic performance index, *Computers and Chemical Engineering*, **1987**, *11*, 77.

152 Zheng, A., A computationally efficient nonlinear MPC algorithm. In: *Proceedings of American Control Conference*, pp. 623–1627, **1997**.

153 Bequette, B. W., Nonlinear Control of Chemical Processes – A Review, *Ind. Eng. Chem. Res.*, **1991**, *30*, 1391–1413.

154 Garcia, C. E., Quadratic dynamic matrix control of nonlinear processes. An application to a batch reaction process, AIChE Annual Meeting, San Francisco, California, **1984**.

155 Brengel, D. D., Geider, W. D., Multistep nonlinear predictive controller, *Ind. Eng. Chem. Res.*, **1989**, *20*, 1812–1822.

156 Li, W. C., Biegler, L. T., Newton-type controllers for constrained nonlinear processes with uncertainty, *Ind. Eng. Chem. Res.*, **1990**, *29*, 2647–1657.

157 Gattu, G., Zafiriou, E., Nonlinear quadratic dynamic matrix control with state estimation, *Ind. End. Chem. Res.*, **1992**, *31*, 1096–1104, 1992.

158 Lee, J. H., Ricker, N. L., Extended Kalman filter based nonlinear model predictive control, *Ind. Eng. Chem. Res.*, **1994**, *33*, 1530–1541.

159 Bequette, B. W., Nonlinear predictive control using multi-rate sampling, *Can. J. Chem. Eng.*, **1991**, *69*, 136–143.

160 Jang, S.-S., Joseph, B., Mukai, H., Comparison of two approaches to on-line parameter and state estimation of nonlinear systems, *Ind. Eng. Chem. Process Des. Dev.*, **1986**, *25*, 809–814.

161 Rawlings, J. B., Meadows, E. S., Muske, K. R., Nonlinear model predictive control: A tutorial and survey. In: *Proceedings International Symposium on Advanced Control of Chemical Processes, ADCHEM*, Kyoto, Japan, **1994**.

162 Renfro, J. G., Morshedi, A. M., Asbjornsen, O. A., Simultaneous optimization and solution of systems described by differential/algebraic equations, *Computers and Chemical Engineering*, **1987**, *11*, 503.

163 Bock H. G., Eich, E., Schloder, J. P., Numerical solution of constrained least squares boundary value problems in differential-algebraic equations. In: Strehmel, K. (Ed.), *Numerical Treatment of Differential Equations*, Teubner, **1988**.

164 Patwardhan, A. A., Rawlings, J. B., Edgar, T. F., Nonlinear model predictive control, *Chem. Eng. Commun.*, **1990**, *87*, 123–141.

165 Rao, C. V., Wright, S. J., Rawlings, J. B., Application of interior-point methods to model predictive control. *Journal of Optimization Theory and Applications*, *1998*, *99*, 723–757.

166 Nagy, Z., Agachi, S., Bodizs, L., Adaptive neural network model based nonlinear predictive control of a fluid catalytic cracking unit. In: Pierucci, S. (Ed.), *European Symposium on Computer Aided Process Engineering-10, ESCAPE-10*,

Florence, Italy, 7–11 May, Elsevier Science, pp. 235–240, **2000**.

167 Diehl, M., Bock H. G., Leineweber, D., Schloder, J., Efficient direct multiple shooting in nonlinear model predictive control. In: Keil, F., Mackens, W., Vos, H., Werther, J. (Eds), *Scientific Computing in Chemical Engineering II*, Volume 2, Springer, **1999**.

168 Leineweber D. B., *Efficient reduced SQP methods for the optimization of chemical processes described by large sparse DAE models*, Ph.D. Thesis, University of Heidelberg, **1998**.

169 Agachi, S., Cristea, M., Imre-Lucaci, A., Nagy, Z., Silaghi, L., Industrial system of energy management. In: *12th International Congress of Chemical and Process Engineering, CHISA »96*, Praha, full paper on CD, **1996**.

170 Agachi, S., Nagy, Z., Imre, A., Cristea, M., Szasz, G., Data acquisition and monitoring system of an industrial power station. In: *13th International Congress of Chemical and Process Engineering CHISA »98*, 23–28 August, Praha, full paper on CD, **1998**.

171 Nagy, Z., Agachi, S., Distance learning applied in an introductory data acquisition course. In: *Proceedings, 5th International Conference on Computer Aided Engineering Education, CAEE »99*, Sofia, Bulgaria, pp. 250–254, **1999**.

172 Nagy, Z., Agachi, S., DAQ system for remote monitoring and control of a laboratory chemical plant. In: *Proceedings of the International Conference of "Molecular and Isotopic Processes"*, pp. 129–130, **1999**.

173 Nagy, Z., Agachi, S., Software pentru curs electronic de achizitie de date, cu achizitie de date la distanta si videoconferinta, *Revista Romana de Informatica si Automatica*, **1999**, *9*(3), 65–70.

174 Garcia, C. E., Morari, M., Internal model control-2. Design procedure for multivariable systems, *Ind. Engng Chem. Process Des. Dev.*, **1985**, *24*, 472–484.

175 Garcia, C. E., Morari, M., Internal model control-3. Multivariable control law computation and tuning guidelines, *Ind. Engng Chem. Process Des. Dev.*, **1985**, *24*, 484–494.

176 Shridar, R., Cooper, D. J., A tuning strategy for unconstrained SISO model predictive control, *Ind. Eng. Chem. Res.*, **1997**, *36*, 729–746.

177 Qin S. J., Badgewell, T. A survey of industrial Model Predictive Control technology, *Control Engineering Practice*, **2003**, *11*, 733–764.

178 Semino, D., Scali, C., A method for robust tuning of linear quadratic optimal controllers, *Ind. Eng. Chem. Res.*, **1994**, *33*(4), 889–895.

179 Karla, L., Georgakis, C., The effects of operational characteristics of catalytic cracking reactors on the closed loop performance of linear model predictive controllers, *Computers and Chemical Engineering*, **1996**, *20*(4), 4010–4015.

180 Scattolini, R., Bittanti, S., On the choice of the horizon in long-range predictive control – some simple criteria, *Automatica*, **1990**, *26*(5), 915–917.

181 Cristea, M. V., Agachi, S., Issues on model predictive control tuning, *AQTR 2000 Conference*, Cluj-Napoca, Romania, **2000**.

182 Cristea, M. V., Agachi, S. P., Model predictive control of inferred variables and dynamic sensitivity analysis applied to MPC tuning, *Buletinul Universitatii "Petrol-Gaze" Ploiesti*, **2000**, *52*(1), 52–57.

183 Zafiriou, E., Morari, M., Design of robust digital controllers and sampling-time selection for SISO systems, *Int. J. Control*, **1986**, *44*(3), 711–735.

184 Kalra, L., Georgakis, C., Effect of process nonlinearity on the performance of linear model predictive controllers for the environmentally safe operation of a fluid catalytic cracking unit, *Ind. Eng. Chem. Res.*, **1994**, *33*, 3063–3069.

185 Luenberger, D. J., An introduction to observers, *IEEE Trans. Automat. Contr.*, **1971**, *16*, 596–602.

3
Case Studies

3.1
Productivity Optimization and Nonlinear Model Predictive Control (NMPC) of a PVC Batch Reactor

3.1.1
Introduction

Polyvinyl chloride (PVC) is one of the most important products of the polymer industry. The product is characterized by: K_w, a number related to the average molecular weight; molecular weight distribution (MWD); particle diameter; particle size distribution; and porosity. Thermal properties, stress–strain properties, impact resistance, strength and hardness of films of polymers are all improved by narrowing the MWD [1,2]. Therefore, the typical goal in operating a batch polymerization reactor system is to minimize polydispersity and to achieve a minimum MWD and reaction time [3–8]. Hence, the development of methodology for adjusting the MWD, and reducing the reaction time is desired in many of the polymer industries. A proper fabrication recipe (initial concentration of initiators) or temperature policy would keep all of those parameters which characterize the polymer within certain limits, but with one or two of them possibly being optimized [2,9–12]. In the suspension technology, PVC is obtained in batch reactors at constant temperature. The operation of the process following a certain policy of temperature is a practical problem with important economical effects. It has been shown that for a batch polymerization process, with radical mechanism, optimal policies of temperature can be determined, so that the MWD or the reaction time can be held at a minimum [13]. A genetic algorithm (GA) has been used to determine the optimal fabrication recipe and temperature profile for improving PVC reactor productivity, through minimization of the reaction time and narrowing of the MWD. The results obtained can be used in practice only if an appropriate control technique, which is capable of controlling the process along the obtained optimum temperature profile, exits. In this way, the objectives of this research are on the one hand to optimize the productivity of the reactor, and on the other hand to demonstrate that, with appropriate advanced control algorithms, the temperature control of the reactor can be improved and the results of the optimization can be implemented in practice. These main objectives are summarized in Figure 3.1.

Agachi/Nagy/Cristea/Imre-Lucaci. *Model Based Control*
Copyright © 2006 WILEY-VCH Verlag GmbH & Co. KGaA, Weinheim
ISBN: 3-527-31545-4

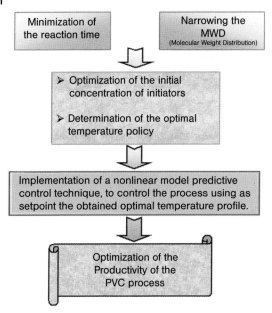

Figure 3.1 Main objectives, and research strategy for the optimization of the PVC reactor.

3.1.2
Dynamic Model of the PVC Batch Reactor

In the simulations performed, two models were used:
- An "external" model, representing the plant, was more complicated and contained the equations for MWD, K_w, porosity, and PVC particle diameter.
- An "inner", simplified model, which was used by the predictive controller to compute the control movements, and which contains only those equations needed to model the dynamics of temperature (the controlled variable), and the function of the cooling medium flow (manipulated variable).

The first model was used in the optimization, and its main structure is illustrated in Figure 3.2.

The process takes place in a batch reactor (Fig. 3.3), following the known mechanism of free-radical polymerization. The process has mainly three stages:

1. Heating of the mass of reaction from room temperature to the polymerization temperature, by introducing steam into the reactor jacket (ca. 1 h).
2. Polymerization is initiated in the first stage and continues for approximately 8–10 h, the heat of reaction being evacuated by the cooling medium.
3. Inhibition of the reaction and cooling of the reactor when a conversion of 90% has been attained. This third stage is not controlled.

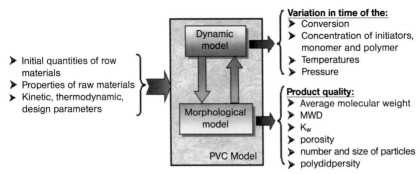

Figure 3.2 The structure of the PVC model used for optimization and simulation as the plant.

During the first stage (heating) the temperature is controlled using a PID controller; its main target is to attain the polymerization temperature as rapidly as possible, but not to exceed it. At 2 °C below the polymerization temperature the steam injection is stopped and control is switched to the cooling system. The main stages and applied control techniques in the reactor operation are illustrated schematically in Figure 3.4.

3.1.2.1 The Complex Analytical Model of the PVC Reactor

3.1.2.1.1 The Physico-Chemical Properties of the Mass of Reaction
The mass of reaction consists of water, vinyl chloride (VC) and PVC as the main components; thus, the physico-chemical properties depend mainly on the aforementioned components:

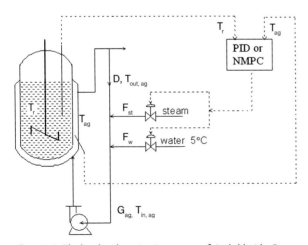

Figure 3.3 The batch polymerization reactor of vinyl chloride. For details of terms, see Section 3.1.6.

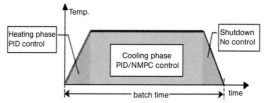

Figure 3.4 The three main stages in the PVC batch reactor operation.

- the density of the mixture

$$\varrho_r = \frac{1}{\sum \frac{x_i}{\varrho_i}}$$

(3.1)

- the specific heat

$$C_p = \sum x_i \cdot c_{pi}$$

(3.2)

where:

$$x_i = \frac{w_a \cdot \varrho_r}{m_r \cdot \varrho_a}$$

$$x_p = \frac{m^0 \cdot \xi \cdot \varrho_p}{m_r \cdot \varrho_p}$$

$$x_{VC} = \frac{(1 - \xi) \cdot m^0 \cdot \varrho_r}{m_r \cdot \varrho_{VC}}$$

Then, Eq. (3.2) becomes:

$$\varrho_r = \frac{m_r}{\frac{m_a}{\varrho_a} + \frac{m^0 \cdot \xi}{\varrho_p} + \frac{m^0 \cdot (1 - \xi)}{\varrho_{VC}}}$$

(3.3)

- the thermal conductivity of the mixture is expressed using Maxwell's equation, written for the electrical conductivity of discontinuous media:

$$\lambda_r = \lambda_1 \cdot \frac{2 \cdot \lambda_1 + \lambda_p - 2 \cdot \xi \cdot (\lambda_1 - \lambda_p)}{2 \cdot \lambda_1 + \lambda_p + \xi \cdot (\lambda_1 - \lambda_p)}$$

(3.4)

The experimental data deviate from the computed values only in the range of 5% for $\xi = 0,01$; for $\xi = 0,3$, the deviations are about 30%. In the case of the practical application followed, conversion has greater values around 0.75; hence, we preferred to use the relationship (3.5), which is simple and additive in nature:

$$\lambda_r = \frac{w_a \cdot \lambda_a + m_{VC} \cdot (1 - \xi) \cdot \lambda_{VC} + m_{VC} \cdot \xi \cdot \lambda_p}{m_r}$$

(3.5)

- the viscosity of the liquid is:

$$\eta_1 = \frac{x_a + x_{VC}}{\dfrac{x_{VC}}{\eta_{VC}} + \dfrac{x_a}{\eta_a}} \tag{3.6}$$

- and that of the suspension:

$$\eta_s = \eta_1 \cdot \frac{0,59}{\left(\xi_c - x_p\right)^2} \tag{3.7}$$

The dependencies of the main physico-chemical properties on temperature are listed in Table 3.1 [14,15]. Temperatures are expressed in degrees Celsius.

3.1.2.1.2 Kinetics of the Reaction and Balance Equations

The polymerization reactions take place following the well-known free-radical polymerization mechanism presented in Figure 3.5.

Based on the multitude of kinetic studies reported in the literature [16,17], a model was developed based on Ugelstad's relationships [18]. Changing some of the relationships with others from models in which these parts have a better theoretical basis might lead to a more detailed model. Thus, the developed kinetic model is a combination of Ugelstad's model and those developed by Hamielec and Bulle [16].

This model is valid with the condition that the distribution equilibrium of radicals between the two phases is reached quickly, so that there is an interface change of radicals. Thus, the ratio (3.8) is constant:

$$Q = \frac{C_R^P}{C_R^L} = \frac{k_a}{k_d} \tag{3.8}$$

- **Initiation**

$I \longrightarrow 2R^\cdot$ with rate constant k_d

$R^\cdot + M \longrightarrow P_1^\cdot$ with rate constant k_i

- **Propagation**

$P_n^\cdot + M \longrightarrow P_{n+1}^\cdot$ with rate constant k_p

- **Chain Transfer to Monomer**

$P_n^\cdot + M \longrightarrow M_n + P_1^\cdot$ with rate constant k_{trm}

- **Chain Transfer to Solvent**

$P_n^\cdot + S \longrightarrow M_n + S_i^\cdot$ with rate constant k_{trs}

- **Dead Polymer Termination**

$P_n^\cdot + P_m^\cdot \longrightarrow$ termination with rate constant k_t

Figure 3.5 The mechanism of free-radical polymerization.

Table 3.1 The physico-chemical properties of the main components of the mass of reaction.

Component	Properties	Expression	UM
Water	ϱ	$1001.23 \cdot (1 + 7.95 \cdot 10^{-5 \cdot t} + 3.74 \cdot 10^{-6 \cdot t2})$	kg m^{-3}
	c_p	$4.19 - 1.048 \cdot 10^{-3 \cdot t} + 1.31 \cdot 10^{-5 \cdot t2}$	kJ (kg·K)$^{-1}$
	λ	$1.999 \cdot (1 + 0.003 \cdot t)$	kJ (m·h·K)$^{-1}$
	η	$2.78 \cdot 10^{-5} \exp[528/(126.52 + t)]$	kg (m·s)$^{-1}$
VC	ϱ	$947.1 - 1.746 \cdot t - 3.24 \cdot 10^{-3 \cdot t2}$	kg m^{-3}
	c_p	$1.2456 + 0.0054 \cdot t$	kJ (kg·K)$^{-1}$
	λ	$0.4806 - 12.351 \cdot 10^{4 \cdot t}$	kJ (m·h·K)$^{-1}$
	η	$1.352 \cdot 10^{-5} \exp[745/(273.15 + t)]$	kg (m·s)$^{-1}$
PVC	ϱ	1400	kg m^{-3}
	c_p	$0.5995 + 0.0025 \cdot t$	kJ (kg·K)$^{-1}$
	η	0.59	kJ (m·h·K)$^{-1}$

UM = unit of measurement.

In this case, the quasi-steady-state condition for the number of radicals is written as:

$$\frac{dn_R^L}{dt} = 2 \cdot k_i \cdot n_i - k_a \cdot C_R^L + k_d \cdot C_R^P - 2 \cdot k_{tL} \cdot C_R^{L2} \cdot V_L = 0 \tag{3.9}$$

$$\frac{dn_R^P}{dt} = 2 \cdot k_i \cdot n_i + k_a \cdot C_R^L - k_d \cdot C_R^P - 2 \cdot k_{tP} \cdot C_R^{P2} \cdot V_P = 0 \tag{3.10}$$

From Eqs. (3.8) to (3.10) we can obtain:

$$C_R^L = \left[\frac{2 \cdot k_i \cdot n_i}{k_{tL} \cdot V_L + Q^2 \cdot k_{tP} \cdot V_P} \right]^{\frac{1}{2}} \tag{3.11}$$

Taking into consideration that:

$$-\frac{dn_M}{dt} = k_C \cdot C_M^L \cdot C_R^L \cdot V_L + k_P \cdot C_M^P \cdot C_R^P \cdot V_P \tag{3.12}$$

one obtains:

$$-\frac{dn_M}{dt} = k_P \cdot \sqrt{\frac{2 \cdot k_i \cdot n_i}{k_{tL} \cdot V_L + Q^2 \cdot k_{tP} \cdot V_P}} \cdot (n_M^L + Q \cdot n_M^P) \tag{3.13}$$

where:

$$C_R^L = \sqrt{\frac{k_{n1} \cdot n_{I1} + k_{n2} \cdot n_{I2}}{k_{tL} \cdot V_L + k_{tP} \cdot V_P \cdot Q^2}} \quad \text{for } \xi > 0.01 \tag{3.14}$$

$$C_R^L = \sqrt{\frac{k_{n1} \cdot n_{I1} + k_{n2} \cdot n_{I2}}{k_{tL} \cdot V_L}} \text{ for } \xi \leq 0.01 \tag{3.15}$$

$$C_R^P = Q \cdot C_R^L \text{ for } \xi > 0.01 \tag{3.16}$$

$$C_R^P = 0 \text{ for } \xi \leq 0.01 \tag{3.17}$$

The volumes of the two phases are computed using the following equations:

$$V_L = \frac{n_M^L}{\varrho_{VC}} = \frac{m^0 - m_P - m_M^P}{\varrho_{VC}} = \frac{m^0 \cdot (1 - \xi - A \cdot \xi)}{\varrho_{VC}} = V_0 \cdot (1 - \xi - A \cdot \xi) \tag{3.18}$$

$$V_P = \frac{m_{PVC}}{\varrho_{PVC}} + \frac{m_M^P}{\varrho_{VC}} = \frac{m^0 \cdot \xi}{\varrho_{PVC}} + \frac{A \cdot m^0 \cdot \xi}{\varrho_{VC}} = V_0 \cdot \xi \cdot \left(A + \frac{\varrho_{VC}}{\varrho_{PVC}} \right) \tag{3.19}$$

Substituting in Eq. (3.13), one can obtain the expression giving the evolution of the conversion with time:

$$\frac{d\xi}{dt} = k_p \sqrt{\frac{k_i \cdot n_i}{k_{tL} \cdot V_0 \cdot (1 - \xi - A \cdot \xi) + Q^2 \cdot k_{tP} \cdot V_0 \cdot \xi \cdot \left(A + \frac{\varrho_{VC}}{\varrho_{PVC}} \right)}} (1 - \xi - A \cdot \xi + Q \cdot A \cdot \xi) \tag{3.20}$$

The moment when the free monomer has been consumed is marked by the critical conversion ξ_C, determined from the condition $V_L = 0$.

Variation of the number of the moles of initiator is described by:

$$\frac{dn_i}{dt} = -k_i \cdot n_i \tag{3.21}$$

At conversions higher than the critical one (ξ_C), the liquid phase disappears and the reaction takes place only in the polymer phase. Hence, the kinetic equation becomes:

$$\frac{d\xi}{dt} = k_p \cdot \left(\frac{k_i \cdot n_i}{k_{tP} \cdot V_P} \right)^{\frac{1}{2}} \cdot \frac{n_M^P}{n_M^0} \tag{3.22}$$

or

$$\frac{d\xi}{dt} = k_p \cdot \left(\frac{k_i \cdot n_i}{k_{tP} \cdot V_P} \right)^{\frac{1}{2}} \cdot (1 - \xi) \tag{3.23}$$

where V_P, the volume of polymer can be computed:

$$V_P = \frac{m_{PVC}}{\varrho_P} + \frac{m_M^P}{\varrho_{VC}} = \frac{m^0 \cdot \xi}{\varrho_P} + \frac{(1 - \xi) \cdot m^0}{\varrho_{VC}} \tag{3.24}$$

or, in a simpler form:

$$V_P = V_0 \cdot \left(1 - \xi + \frac{\varrho_{VC}}{\varrho_P} \cdot \xi\right) \tag{3.25}$$

When a mixture of initiators is used, the term $k_i \, n_i$ is used rather than $\sum k_i \, n_i$. Usually, as noted previously, two initiators are used (fast and slow) in order to obtain a rate of reaction which is more uniform along the batch.

The balance on monomer is given in Eq. (3.26):

$$m_0 = m_1 + m_P + m_G + w_P + m_m \tag{3.26}$$

These terms are computed as follows:

$$w_P = m^0 \cdot \xi \tag{3.27}$$

$$m_w = K_{sol} \cdot w_a \tag{3.28}$$

where K_{sol}, the constant of solubility of monomer in water is:

$$K_{sol} = 0,0472 - \frac{11,6}{T_r} \tag{3.29}$$

$$m_P = A \cdot w_P = A \cdot m^0 \cdot \xi \tag{3.30}$$

The constant A has the following dependence on temperature:

$$A = 0,18856 - 0,0044593 \cdot T_r + 0,0001413 \cdot T_r^2 \tag{3.31}$$

For conversions greater than ξ_C, A is obtained by solving the Flory–Huggins equation:

$$\ln\frac{p_m}{p_m^0} = \ln(1 - \varphi) + \left(1 - \frac{1}{n}\right) \cdot \varphi + \chi \cdot \varphi^2 \tag{3.32}$$

where:

$$\chi = \frac{1286,4}{T_r} - 3.02 \tag{3.33}$$

Because n is a large number (≈ 100), $1 - \frac{1}{n} \approx 1$:

$$A = \frac{\varrho_m}{\varrho_P} \cdot \left(\frac{1}{\varphi} - 1\right) \tag{3.34}$$

and then:

$$\varrho = \frac{1}{1 + A \cdot \frac{\varrho_P}{\varrho_m}} \tag{3.35}$$

Because there exist free monomer still ($\xi < \xi_C$):

$$p_m = p_m^0 = 12722 \cdot e^{\left(\frac{-2411.7}{T_r}\right)} \tag{3.36}$$

$$m_G = V_G \cdot \frac{p_m^0 \cdot M_{VC}}{R \cdot T_r} = V_G \cdot \varrho_G \tag{3.37}$$

$$V_G = V_R - V_u \tag{3.38}$$

$$V_u = \frac{w_a}{\varrho_a} + \frac{w_P}{\varrho_P} + \frac{m_P}{\varrho_m} + \frac{m_l}{\varrho_m} \tag{3.39}$$

$$m_G = \varrho_G \cdot \left(V_R - \frac{w_a}{\varrho_a} - \frac{w_P}{\varrho_P} - \frac{m_P}{\varrho_m} - \frac{m_l}{\varrho_m} \right) \tag{3.40}$$

From Eqs. (3.23) to (3.25), (3.27), and (3.39) one obtains:

$$m_L = \frac{m^0 - \varrho_G V_R - \left[\left(1 - \frac{\varrho_G}{\varrho_m} \right) K_{sol} - \frac{\varrho_G}{\varrho_a} \right] w_a - \left[\left(1 - \frac{\varrho_G}{\varrho_P} \right) + A \left(1 - \frac{\varrho_G}{\varrho_m} \right) \right] m^0 \xi}{1 - \frac{\varrho_G}{\varrho_m}} \tag{3.41}$$

The volume of liquid monomer, being:

$$V_L = \frac{m_L}{\varrho_m} \tag{3.42}$$

At $V_L = 0$, one computes the critical conversion ξ_C:

$$\xi_c = \frac{m^0 - \varrho_G \cdot V_R - \left[\left(1 - \frac{\varrho_G}{\varrho_m} \right) \cdot K_{sol} - \frac{\varrho_G}{\varrho_a} \right] \cdot w_a}{m^0 \cdot \left[\left(1 - \frac{\varrho_G}{\varrho_P} \right) + A \cdot \left(1 - \frac{\varrho_G}{\varrho_m} \right) \right]} \tag{3.43}$$

For ξ greater than ξ_C, the balance on monomer is written:

$$m^0 = m_G + m_w + m_P + w_P \tag{3.44}$$

- where the mass of the monomer solved in water:

$$m_w = K_{sol} \cdot w_a \cdot \frac{p_m}{p_m^0} \tag{3.45}$$

- the mass of polymer:

$$w_P = m^0 \cdot \xi \tag{3.46}$$

- the mass of monomer solved in polymer:

$$m_P = A \cdot m^0 \cdot \xi \tag{3.47}$$

- the mass of monomer in vapor phase:

$$m_G = (V_R - V_u) \cdot \frac{p_m \cdot M_m}{R \cdot T_r} \tag{3.48}$$

$$m_G = \frac{p_m \cdot M_m}{R \cdot T} \cdot \left[V_R - \frac{w_a}{\varrho_a} - \frac{w_P}{\varrho_P} - \frac{M_P}{\varrho_m} \right] \tag{3.49}$$

Substituting in Eq. (3.44) all of the computed terms in Eqs. (3.45) to (3.49), A becomes:

$$A = \frac{\left[1 - \left(1 - \frac{\varrho_G}{\varrho_P} \right) \cdot \xi \right] \cdot m^0 - \left[\left(1 - \frac{\varrho_G}{\varrho_m} \right) \cdot K_{sol} \cdot \frac{p_m}{p_m^0} - \frac{\varrho_G}{\varrho_m} \right] \cdot w_a - \varrho_G \cdot V_R}{m^0 \cdot \xi \cdot \left(1 - \frac{\varrho_G}{\varrho_m} \right)} \tag{3.50}$$

3.1.2.1.3 Calculation of Pressure in the Reactor

At conversions less than ξ_C, there exists a liquid phase of the monomer, and the partial pressure is then equal to the vapor pressure. The total pressure of the system is:

$$p_t = p_m^0 + p_a^0 \tag{3.51}$$

where:

$$p_a^0 = \frac{1}{760} \cdot \exp \left(47,9 - 4,03 \cdot \ln T_r - \frac{6493,3}{T_r} \right) \tag{3.52}$$

and p_m^0 is given by Eq. (3.36).

At conversions greater than ξ_C, in order to compute the partial pressure of monomer, the mass transfer from vapor space through the liquid towards the reaction zone should be considered:

$$\frac{dn_m}{dt} = k_g \cdot A_{TR} \cdot (p_m - p_m^e) \tag{3.53}$$

$$n_m = \frac{V_G \cdot p_m}{R \cdot T_r} \tag{3.54}$$

From Eqs. (3.50) and (3.51), the result is:

$$\frac{dp_m}{dt} = \frac{R \cdot T_r}{V_G} \cdot k_g \cdot A_{TR} \cdot (p_m - p_m^e) \tag{3.55}$$

The equilibrium pressure is calculated using the Flory–Huggins equation written in another form:

$$p_m^e = p_m^0 \cdot \exp \left[\varphi + \chi \cdot \varphi^2 \right] \cdot (1 - \varphi) \tag{3.56}$$

The total pressure of the reactor is then computed with Eq. (3.51).

3.1.2.1.4 Determination of the Average Molecular Weight

The weight fraction of the polymer of a chain length r,

$$w_r = g_1 \cdot w_{r1} + g_P \cdot w_{rP} \tag{3.57}$$

$$w_r = \tau^2 \cdot r \cdot e^{(-\tau * r)} \tag{3.58}$$

where:

$$\tau = \frac{r_{trm} + r_t}{r_P} \tag{3.59}$$

Substituting the expressions of rate of reaction:

$$\tau_1 = \frac{k_{trm}}{k_P} + \frac{k_{t1} \cdot C_R^L}{k_P \cdot C_M^L} \cdot \frac{\lambda_L + 1}{2}$$
$$\tau_P = k_{tr} \cdot \frac{m}{k_P} + \frac{k_{tP} \cdot C_R^P}{k_P \cdot C_M^P} \cdot \frac{\lambda_P + 1}{2} \tag{3.60}$$

where $\lambda_{L,P} \in [0, 1]$,
$\quad\quad \lambda = 1$ for 100 % termination by disproportioning, and
$\quad\quad \lambda = 0$ for 100 % termination by reunification of two radicals.

When $\xi > \xi_C$ we have only:

$$\tau_P = \frac{k_{trm}}{k_P} + \frac{k_{tP} \cdot C_R}{k_P \cdot C_M^P} \cdot \frac{\lambda_P + 1}{2} \tag{3.61}$$

$$C_M^L = \frac{\varrho_m}{m_{VC}} \tag{3.62}$$

$$C_M^P = \frac{m_P}{V_P \cdot m_{VC}} \tag{3.63}$$

The weight fractions of the polymer in the two phases are:

$$g_l = \frac{r_P^l}{r_P} \quad \text{and} \quad g_P = \frac{r_P^P}{r_P} \tag{3.64}$$

where:

$$r_P^l = k_P \cdot C_R^L \cdot \frac{m_l}{m^0} \tag{3.65}$$

$$r_P^P = k_P \cdot C_R^P \cdot \frac{m_P}{m^0} \tag{3.66}$$

It is then possible to compute the weight distribution:

$$\mu_{-1} = \int_0^\infty \frac{w_r}{r} \cdot dr = g_l \cdot \tau_l + g_P \cdot \tau_P$$

$$\mu_0 = \int_0^\infty w_r \cdot dr = 1 \tag{3.67}$$

$$\mu_1 = \int_0^\infty r \cdot w_r \cdot dr = 2 \cdot \left(\frac{g_l}{\tau_l} + \frac{g_P}{\tau_P} \right)$$

- The number-average molecular weight is:

$$\bar{M}_n = \frac{\mu_0}{\mu_{-1}} \cdot m_m = \frac{m_m}{g_l \cdot \tau_l + g_P \cdot \tau_P} \tag{3.68}$$

- The weight-average molecular weight:

$$\bar{M}_w = \frac{\mu_1}{\mu_0} \cdot m_m = 2 \cdot \left(\frac{g_l}{\tau_l} + \frac{g_P}{\tau_P} \right) \cdot m_m \tag{3.69}$$

The means are computed at small time intervals during which the polymerization speed can be considered constant and, ultimately, the overall mean is calculated:

$$\bar{M} = \frac{1}{\xi_f} \cdot \sum_{i=1}^{n} (\xi_{i+1} - \xi_i) \cdot M_i \tag{3.70}$$

The parameter K_w is computed via empirical equation from M_n:

$$K_w = 31 + 6.222 \cdot 10^{-6} \cdot M_n \tag{3.71}$$

3.1.2.1.5 Heat Balance

The equations describing heat balance are mainly those written for the reaction space and jacket:

$$\frac{d}{dt} \left(m_r \cdot c_{pr} \cdot T_r \right) = Q_r - Q_T \tag{3.72}$$

$$\frac{d}{dt} \left(V_m \cdot \varrho_{ag} \cdot c_{pag} \cdot T_{ag} \right) = Q_T + G_{ag} \cdot c_{pag} \cdot \left(T_{iag} - T_{eag} \right) \tag{3.73}$$

$$Q_r = -\Delta H_r \cdot m^0 \cdot \frac{d\xi}{dt} \tag{3.74}$$

$$Q_T = K_T \cdot A_T \cdot \Delta T_{med} \tag{3.75}$$

Taking into account a fraction D of the cooling water at the exit of the jacket is recycled:

$$G_{ag} = D + F_A + G_{ab} \tag{3.76}$$

$$G_{ag} \cdot T_{iag} = D \cdot T_{eag} + F_A \cdot T_a + \frac{i_{ab}}{c_{pa}} \cdot G_{ab} \tag{3.77}$$

The overall heat transfer coefficient K_T is calculated from the equations:

$$K_T = \frac{1}{\frac{1}{\alpha_i} + \sum r_i + \frac{1}{\alpha_e}} \tag{3.78}$$

$$\alpha_i = \frac{Nu_i \cdot \lambda_r}{D_r}$$
$$\alpha_e = \frac{Nu_e \cdot \lambda_{ag}}{d_e} \tag{3.79}$$

$$Nu_i = 0.38 \cdot Re_i^{0.67} \cdot Pr_i^{0.33} \cdot \frac{d_{ag}}{d_r} \tag{3.80}$$

$$Re_i = \frac{D_{ag}^2 \cdot n_{rot} \cdot \varrho_r}{\eta_s} \tag{3.81}$$

For the spiral jacket, the Nusselt number is calculated with the formula for turbulent flow fully developed:

$$Nu_e = 0.021 \cdot Re_e^{0.8} \cdot Pr^{0.63} \cdot \left(\frac{Pr}{Pr_p}\right)^{0.25} \tag{3.82}$$

where:

$$\left(\frac{Pr}{Pr_p}\right)^{0.25} = \begin{cases} 1 & \text{for cooling} \\ 0.93 & \text{for heating} \end{cases}$$

$$Re = \frac{d_e \cdot \varrho_{ag} \cdot v}{\eta_{ag}}$$
$$Pr = \frac{c_{pa} \cdot \eta_a}{\lambda_{ag}} \tag{3.83}$$

The thermal resistances considered are those of the wall, of the inner polymer crust, and of the outer deposit from water:

$$\sum r_i = r_w + r_{dep} + r_{cr} \tag{3.84}$$

$$r_{cr} = \frac{\delta_{cr}}{\lambda_{VC}} \tag{3.85}$$

The thickness of the crust is approximated as a function of time:

$$\delta_{cr} = 1.4 \cdot 10^{-7} \cdot t \tag{3.86}$$

During the batch, K_T decreases from values around 710 $\dfrac{W}{m^2 \cdot K}$ to values of 450 $\dfrac{W}{m^2 \cdot K}$. This shows that, towards the end of reaction, the system has a more difficult task to remove the heat. It is possible then, that the peak rate of reaction can be predicted and the system be cooled in advance.

3.1.2.2 Morphological Model

During polymerization of the VC, a sequence of particle formation and congestion takes place inside the VC drop, at the microscopic level.

In order to explain the formation of the PVC particle structure, the following terminology is used:

- *Microparticles*: size 10 to 20 nm; these are aggregations of polymer chains or germs of polymerization.
- *Domains*: size 100 to 200 nm; these are aggregations of microparticles or nuclides of primary particles.
- *Primary particles*: size 600 to 800 nm; these are aggregations of domains.
- *Conglomerate*: size 1000 to 10 000 nm; these are aggregations of primary particles.
- *Subgranules*: size 10 000 to 15 000 nm; the PVC granules formed from one VC drop.
- *Granules*: size 50 000 to 250 000; PVC granules formed from aggregations of VC drops.

The stages of the PVC morphogenesis are:

- The formation of macromolecules and dispersion of chains with the formation of domains or microparticles ($\xi = 0$).
- Aggregation and growing until the formation of primary particles ($\xi \leq \xi_1$).
- Formation and consolidation of the agglomerations of primary particles ($\xi_1 < \xi < \xi_2$).
- Growing of agglomerations up to the steric constraint ($\xi = \xi_3$).
- Growing of agglomerations until the pressure begins to decrease ($\xi < \xi_c$).
- Formation of structure of the subgranules and granules until the final conversion $\left(\xi_c \leq \xi \leq \xi_f\right)$.

3.1.2.2.1 Stage 1

The polymer chains, initiated by the radical mechanism, grow until a length of 25 to 32 units is reached. In electronic microscopy studies it was shown that in some cases, the polymer chains might reach a length of 200–500 monomer units.

The number and diameter of the polymerization germs can be obtained using Eqs. (3.87) to (3.90):

$$N = \xi_{cr1} \cdot \frac{\varrho_m \cdot A_V \cdot V_R}{P_{n,k} \cdot M_{CV} \cdot z_K} \tag{3.87}$$

where the volume of the reaction medium V_R is:

$$V_R = \frac{m_m^0}{\bar{\varrho}} \tag{3.88}$$

and the density of the medium is computed with the following equation:

$$\bar{\varrho} = \frac{1}{\dfrac{\xi}{\varrho_p} + \dfrac{1-\xi}{\varrho_m}} \tag{3.89}$$

The diameter of the microparticles is:

$$d = \sqrt[3]{\frac{6 \cdot V_G}{\pi \cdot N}} \tag{3.90}$$

The volume and the density of the gel phase V_G are:

$$V_G = \frac{\xi \cdot m_m^0}{\xi_c \cdot \varrho_G} \tag{3.91}$$

$$\varrho_G = \frac{1}{\dfrac{\xi_c}{\varrho_p} + \dfrac{1-\xi_c}{\varrho_m}} \tag{3.92}$$

The critical conversion of saturation ξ_{cr1} is difficult to determine. It is defined as the conversion at which the saturation concentration of the polymer germs is reached. When this concentration is reached, it is considered that every z_K polymer chain with a length of $P_{n,k}$ will aggregate spontaneously, leading to a microparticle. At moment t_k, corresponding to this conversion, the diameter of the particles is given by Eq. (3.90). Between the conversions ξ_{cr1} and ξ_1, the dispersibility of the microparticles has a constant value (i.e., $N_1 = N = $ constant.). When conversion ξ_1 is reached, the formation of the first conglomeration of microparticles is possible due to collisions which take place.

3.1.2.2.2 Stages 2 and 3

Once conversion ξ_1 is achieved, N decreases and the diameter of particles increases due to polymerization, conglomeration, and inhibition with monomer. The average distance $\Delta\bar{x}_0$ between the surfaces of particles is computed using the following relationship:

$$\Delta\bar{x}_0 = \left(\frac{V_R}{N}\right)^{\frac{1}{3}} - d \tag{3.93}$$

The time interval between collisions is:

$$\Delta t = \frac{\Delta \bar{x}_0}{2 \cdot D} \tag{3.94}$$

where D is the coefficient of diffusion $\left[\frac{m^2}{s}\right]$ given by the following relationship:

$$D = \frac{R \cdot T_r}{3 \cdot A_V \cdot \pi \cdot \eta \cdot d} \tag{3.95}$$

Considering the conglomeration coefficient K:

$$K = \frac{3 \cdot A_V \cdot \pi}{2 \cdot R} \tag{3.96}$$

one obtains:

$$\Delta t = K \cdot \frac{\Delta \bar{x}_0^2 \cdot d \cdot \eta}{T_r} \tag{3.97}$$

The conglomeration process was discretized by considering the following assumptions:
- not every statistical collision of two particles lead to conglomeration;
- every z_1 particle constitutes a larger particle which can conglomerate later; and
- after a step of conglomeration the number and the diameter of particles become

$$N = \frac{N}{z_1} \tag{3.98}$$

$$d = \left(\frac{z_1}{1 - \varepsilon_1}\right)^{\frac{1}{3}} \cdot d \tag{3.99}$$

where ε_1 is the porosity in the conglomerate.

The number of particles decreases not only due to the conglomeration of each z_1 particle, but also because of the formation of a film of protection. This film is formed due to the centrifugal forces caused by rotation of the drops. The force balance written for a spherical particle inside a VC drop is:

$$F_r = F_d \tag{3.100}$$

where F_r – is the resistance of the medium $[N]$;, and F_d – is the moving force $[N]$.
Writing explicitly the forces, which act upon a particle, Eq. (3.100) becomes:

$$\lambda \cdot \varrho_m \cdot v \cdot d^2 \cdot \frac{\pi}{8} = a \cdot d^3 \cdot \varrho_G \cdot \frac{\pi}{6} - a \cdot d^3 \cdot \varrho_m \cdot \frac{\pi}{6} \tag{3.101}$$

where

$$v = \frac{dr}{dt} - \text{the moving velocity} \ \left[\frac{m}{s}\right]$$

$$a = r\omega_T^2 - \text{acceleration} \ \left[\frac{m}{s^2}\right]$$

Using the following relationships:

$$\lambda = \frac{24}{Re} \tag{3.102}$$

$$Re = \frac{v \cdot d \cdot \varrho_m}{\eta_m} \tag{3.103}$$

where: $\lambda-$ is the frictional coefficient,

η_m- is the viscosity of VC $\left[\dfrac{kg}{m \cdot s}\right]$; and

ω_T- is the rotational velocity $[s^{-1}]$.

One can rewrite Eq. (3.101) in the following form:

$$v = \frac{dr}{dt} = r \cdot \omega_T^2 \cdot d^2 \cdot \frac{(\varrho_G - \varrho_m)}{18 \cdot \eta_m} \tag{3.104}$$

Integrating Eq. (3.104), one obtains:

$$h \cdot \frac{r}{r_p} = -\frac{\omega_T^2 \cdot d^2 \cdot \Delta t_{rot} \cdot (\varrho_G - \varrho_m)}{18 \cdot \eta_m} \tag{3.105}$$

By considering the relationships below:

$$\Delta t_{rot} = \tau \cdot \Delta t \tag{3.106}$$

$$K_{rot} = \frac{\omega_T^2 \cdot \tau \cdot (\varrho_G - \varrho_m)}{18 \cdot \eta_m} \tag{3.107}$$

the following equation can be derived:

$$r = r_p \cdot e^{(-K_{rot} \cdot d^2 \cdot \Delta t)} \tag{3.108}$$

The number of particles participating in the formation of the film of protection is:

$$\Delta N = N - N \cdot \frac{V}{V_p} = N \cdot \left[1 - \left(\frac{r}{r_p}\right)^3\right] \tag{3.109}$$

or

$$\Delta N = N \cdot \left[1 - e^{(K_{rot} \cdot d^2 \cdot \Delta t)}\right]^3 \tag{3.110}$$

3.1.2.2.3 Stage 4

One characteristic of the model is that, initially, the conglomeration steps are very small, but subsequently Δt increases rapidly. Thus, particles grow until the steric limitation, when Δx_0 becomes negative and $\xi = \xi_2$ and $d = d_2$, respectively, correspond to the last positive value of Δx_0.

From the condition $\Delta x_0 = 0$ one can obtain ξ_3 and d_3:

$$\xi_3 = \frac{\dfrac{\pi}{6} \cdot \left(\dfrac{1}{\varrho_m} - N \cdot \dfrac{d_2^3}{m_m^0} \right) + \dfrac{\xi_2}{\xi_c \cdot \varrho_G}}{\dfrac{\pi}{6} \cdot \left(\dfrac{1}{\varrho_m} - \dfrac{1}{\varrho_P} \right) + \dfrac{1}{\xi_c \cdot \varrho_G}}$$

(3.111)

$$d_3 = \sqrt[3]{\frac{V_R}{N}}$$

(3.112)

3.1.2.2.4 Stage 5

The growing of conglomerates, due to increased conversion and repeated conglomeration, leads to rupture of the membrane and the formation of subgranules of irregular shape. The volume, V, of these subgranules can be obtained using Eq. (3.113):

$$V = N \cdot \left[\frac{\pi \cdot d^3}{6} - z_B \cdot \frac{\pi}{12} \cdot (d - d_3)^2 \cdot \left(d + \frac{d_3}{2} \right) \right]$$

(3.113)

where $z_B = 6$—is the coordination number.

3.1.2.2.5 Stage 6

Once the critical conversion ξ_c is reached, the volume of the conglomerates begins to decrease, because the monomer inside it is consumed. In this case, the volume of the particles is computed as follows:

$$V = V_0 + m_m^0 \cdot (\xi - \xi_c) \cdot \left(\frac{1}{\varrho_p} - \frac{1}{\varrho_m} \right)$$

(3.114)

where V_0— is the volume of the subgranule at the critical conversion ξ_c

The decrease in particle diameter due to volume contraction can be obtained using the as follows:

$$\Delta V = V' - V_0$$

(3.115)

where the final diameter of the subgranule is the diameter of the sphere with volume V'.

3.1.2.3 The Simplified Dynamic Analytical Model of the PVC Reactor

In order to improve the real-time feasibility of on-line optimization problems associated with analytical model-based NMPC strategies, only the most important, control-relevant modeling equations were retained in the model used to predict the controller, and these described the dynamic behavior of the reactor with appropriate accuracy. Consequently, the morphological model is completely omitted, together with some of the equations describing average molecular weights and distribution. All of the modeling equations used in this simplified model are described in the complex analytical model, but for consistency, these equations are presented here briefly:

- Physico-chemical properties

$$\varrho_r = \frac{m_r}{\dfrac{m_w}{\varrho_w} + \dfrac{m^0 \cdot \xi}{\varrho_{PVC}} + \dfrac{m^0 \cdot (1 - \xi)}{\varrho_{VC}}} \tag{3.116}$$

$$c_{p,r} = w_w \cdot c_{p,w} + w_{PVC} \cdot c_{p,PVC} + w_{VC} \cdot c_{p,VC} \tag{3.117}$$

$$\lambda_r = \frac{m_w \cdot \lambda_w + m^0 \cdot \xi \cdot \lambda_{PVC} + m^0 \cdot (1 - \xi) \cdot \lambda_{VC}}{m_r} \tag{3.118}$$

$$K_{sol} = 0.0472 - \frac{11.6}{T_r} \tag{3.119}$$

$$\Delta H_r = -1.64 \cdot 10^6 + 157.2 \cdot T_r \tag{3.120}$$

The physico-chemical properties of the main components of the mass of reaction, as a function of the temperature, are listed in Table 3.1 [14,15].

- Kinetic equations; two initiators are used: a fast one (peroxydicarbonic acid bis (1-ethylhexyl) ester; DEHPC), and a slow one (bis (1- oxododecyl) peroxide; LPO).

$$k_{I1} = 4.4 \cdot 10^{17} \cdot e^{\frac{-29833}{R \cdot T_r}} \tag{3.121}$$

$$k_{I2} = 8.5 \cdot 10^{16} \cdot e^{\frac{-30450}{R \cdot T_r}} \tag{3.122}$$

$$k_p = 9.7 \cdot 10^7 \cdot e^{\frac{-3700}{R \cdot T_r}} \tag{3.123}$$

$$k_{tL} = 2.87 \cdot 10^{16} \cdot e^{\left(-5 \cdot \xi - \frac{4200}{R \cdot T_r}\right)} \tag{3.124}$$

$$k_{tP} = 6 \cdot 10^9$$

$$\frac{dn_{I1}}{dt} = -k_{I1} \cdot n_{I1} \tag{3.125}$$

$$\frac{dn_{I2}}{dt} = -k_{I2} \cdot n_{I2} \tag{3.126}$$

for $\xi < \xi_c$

$$A = 0.1856 - 4.456 \cdot 10^{-3} \cdot T_r + 1.413 \cdot 10^{-4} \cdot T_r^2 \tag{3.127}$$

$$\frac{d\xi}{dt} = k_P \sqrt{\frac{k_{I1} n_{I1} + k_{I2} n_{I2}}{k_{tL} V_0 (1 - \xi - A\xi) + Q^2 k_{tP} V_0 \xi \left(A + \frac{\varrho_{VC}}{\varrho_{PVC}} \right)}} \, (1 - \xi - A\xi + QA\xi) \tag{3.128}$$

for $\xi > \xi_c$

$$\varphi = \frac{1}{1 + A \cdot \dfrac{\varrho_{PVC}}{\varrho_{VC}}} \tag{3.129}$$

$$\chi = \frac{1286.4}{T_r} - 3.02 \tag{3.130}$$

$$p_{VC}^0 = 12722 \cdot e^{\left(\frac{-2411.7}{T_r} \right)} \tag{3.131}$$

$$p_{VC}^e = p_{VC}^0 \cdot (1 - \varphi) \cdot e^{(\varphi + \chi \cdot \varphi^2)} \tag{3.132}$$

$$\frac{dp_{VC}}{dt} = \frac{R \cdot T_r}{V_R - V_u} \cdot k_g \cdot A_{mt} \cdot (p_{VC} - p_{VC}^e) \tag{3.133}$$

$$\varrho_G = \frac{p_{VC} \cdot M_{VC}}{R \cdot T_r} \tag{3.134}$$

$$A = \frac{\left[1 - \left(1 - \dfrac{\varrho_G}{\varrho_{PVC}} \right) \cdot \xi \right] \cdot m^0 - \left[\left(1 - \dfrac{\varrho_G}{\varrho_{VC}} \right) \cdot K_{sol} \cdot \dfrac{p_{VC}}{p_{VC}^0} - \dfrac{\varrho_G}{\varrho_{VC}} \right] \cdot m_w - \varrho_G \cdot V_R}{m^0 \cdot \xi \cdot \left(1 - \dfrac{\varrho_G}{\varrho_{VC}} \right)} \tag{3.135}$$

$$\frac{d\xi}{dt} = k_P \cdot \sqrt{\frac{k_{I1} \cdot n_{I1} + k_{I2} \cdot n_{I2}}{k_{tP} \cdot \left(1 - \xi + \dfrac{\varrho_{VC}}{\varrho_{PVC}} \right) \cdot V_0}} \cdot (1 - \xi) \tag{3.136}$$

- Energy balance

$$\frac{d}{dt} (m_r \cdot c_{p,r} \cdot T_r) = -\Delta H_r \cdot m^0 \cdot \frac{d\xi}{dt} - K_T \cdot A_T \cdot (T_r - T_{ag}) \tag{3.137}$$

$$\frac{d}{dt} \left(V_j \cdot \varrho_{ag} \cdot c_{p,ag} \cdot T_{ag} \right) = K_T \cdot A_T \cdot (T_r - T_{ag}) + G_{ag} \cdot c_{p,ag} \left(T_{in,ag} - T_{out,ag} \right) \tag{3.138}$$

$$T_{out,ag} = 2 \cdot T_{ag} - T_{in,ag} \tag{3.139}$$

$$G_{ag} = D + F_w + F_{st} \tag{3.140}$$

$$T_{in,ag} = \frac{D \cdot T_{out,ag} + F_w \cdot T_w + \frac{i_{st}}{c_{p,w}} \cdot F_{st}}{G_{ag}} \tag{3.141}$$

In Eqs. (3.140) and (3.141) during the heating, $F_w = 0$ and $F_{st} = u$ (manipulated variable), but during the reaction, $F_{st} = 0$ and $F_w = u$.

$$K_T = \frac{1}{\dfrac{1}{\alpha_i} + \dfrac{1}{\alpha_e} + \sum r_T} \tag{3.142}$$

$$\alpha_i = \frac{\lambda_r}{d_R} \cdot 0.38 \cdot \left(\frac{d_a^2 \cdot n_{rot} \cdot \varrho_r}{\eta_r}\right)^{0.67} \cdot \left(\frac{c_{p,r} \cdot \eta_r}{\lambda_r}\right)^{0.33} \cdot \frac{d_a}{d_R} \tag{3.143}$$

$$\alpha_e = \frac{\lambda_{ag}}{d_e} \cdot 0.021 \cdot \left(\frac{d_e \cdot \varrho_{ag} \cdot \dfrac{G_{ag}}{A_j}}{\eta_{ag}}\right)^{0.8} \cdot \left(\frac{c_{p,ag} \cdot \eta_{ag}}{\lambda_{ag}}\right)^{0.43} \cdot \left(\frac{Pr}{Pr_p}\right)^{0.25} \tag{3.144}$$

$$\text{where } \left(\frac{Pr}{Pr_p}\right)^{0.25} = \begin{cases} 1 & \text{for cooling} \\ 0.93 & \text{for heating} \end{cases}$$

$$\sum r_T = r_{wall} + r_{cr} \tag{3.145}$$

$$r_{cr} = \frac{\delta_{cr}}{\lambda_{PVC}} \tag{3.146}$$

$$\delta_{cr} = 1.4 \cdot 10^{-7} \cdot t \tag{3.147}$$

With Eq. (3.147), the increase of the polymer deposit is estimated as a function of time.

Thus, the model computes a differential equations system, $x = f(x, u)$ and an algebraic equations one $g(x, u) = 0$, which is much simpler than the detailed model, but it is still complex enough to describe the dynamic behavior of the batch polymerization reactor sufficiently accurately for an appropriate temperature control of this system.

3.1.3
Productivity Optimization of the PVC Batch Reactor

In this section, a brief explanation is provided as to why unconventional techniques should be used to run the process, and why MPC techniques are absolutely necessary.

with:

$$Fitness = \frac{1}{t_r} + \frac{w_{MWD}}{MWD} + \underbrace{\frac{w_{Kw}}{1 + (K_w - 70.60)^2} + \frac{w_{pors}}{1 + (poros - 1.90)^2}}_{Constraints\ introduced\ as\ penalty\ terms} \qquad (3.148)$$

In the optimization, several different weighting factors were tried, and the best results were obtained with $w_{MWD} = 0.01$, $w_{Kw} = 0.006$ and $w_{pors} = 0.001$.

Figure 3.7 (fitness representation) illustrates how the GA functions. While the first generation is evenly distributed on the fitness surface, the members of the last generation (obtained after only 20 generations) are mostly grouped in the neighborhood of the optimal value.

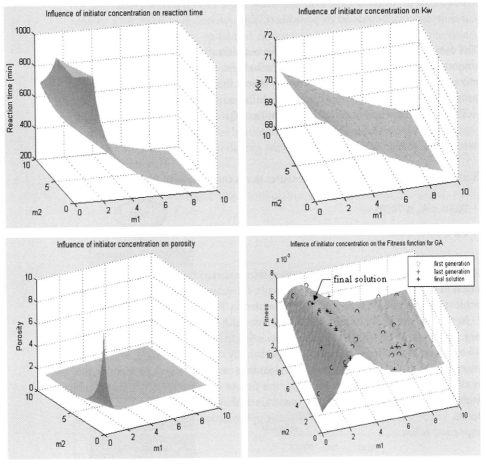

Figure 3.7 Influence of initiator quantities on reaction time, polymer parameters, and fitness.

The final results are presented in Table 3.3. It can be observed that both optimization algorithms reduced the reaction time (t_r) as well as the MWD. The GA outperformed the SQP from the point of view of optimization results. With regard to the speed of convergence expressed as the number of objective function evaluations needed to find a final solution, GA and SQP showed similar performances. However, for this small dimensional problem SQP had a faster convergence in finding the final solution.

Table 3.3 Results obtained with SQP and GA in the determination of the optimal recipe.

Optimi-zation method	Functions evaluated [n]	Reaction temp. in stage 2	Quant. of fast initiator (m1) [kg]	Quant. of slow initiator (m2) [kg]	K_w	Porosity	Reaction time (t_r) [min]	MWD
SQP	212	52 °C, constant	1.6059	5.4433	70.55	1.91	552.44	57 259
GA	320	52 °C, constant	1.5230	6.6542	70.52	1.91	536.20	57 211

When comparing results obtained with the reference values in Table 3.2, it is clear that with SQP the reaction time was reduced by 15 min (2.7%), and the MWD by 0.13%. The results obtained with GA were better: the reaction time was decreased by 31 min (5.5%), and MWD by 0.21%. At an industrial level these results would lead to a significant improvement in the productivity of a PVC plant.

3.1.3.3 Optimization of PVC Reactor Productivity by obtaining an Optimal Temperature Policy

In general, although industrial PVC reactors are operated at a constant temperature, it has been shown that an optimal temperature policy can be obtained for batch polymerization reactors, thereby reducing reaction time and/or MWD. Several different algorithms have been reported which determine the optimal temperature profile. In our studies, it was proposed that a step-shape temperature profile be used, wherein each temperature step was a separate variable in the state space of the optimization problem. The duration of each step (the time for which temperature is considered constant) is chosen carefully, taking into account the settling time of the system. Consequently, steps with duration of 70 min were used.

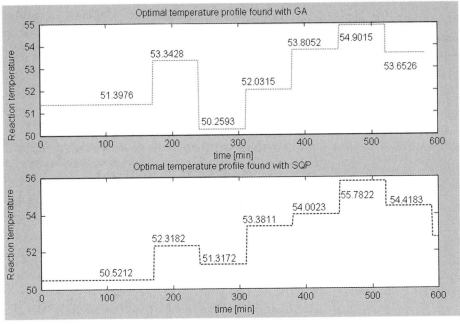

Figure 3.8 Optimal temperature profiles obtained with Sequential Quadratic Programming (SQP) and genetic algorithm (GA).

The optimization problem for the SQP method is expressed as follows:

$$\min_{T_1,T_2,...T_N} \{t_r + w_{MWD} \cdot MWD\} \tag{3.149}$$

subjected to:

$$70.50 \le K_w \le 70.70 \tag{3.150}$$

$$1.8 \le poros \le 2.0 \tag{3.151}$$

In the case of the GA, a fitness function was used with a similar form as Eq. (3.148), though the states were the temperature values in the different time intervals:

$$\min_{T_1,T_2,...T_N} \{Fitness\} \tag{3.152}$$

Optimization was commenced from the optimal recipe found with GA in the previous section; the results are shown in Figure 3.8 and Table 3.4.

Table 3.4 Results obtained with SQP and GA in the determination of the optimal temperature policy.

Optimization method	Functions evaluated [n]	Reaction temp. in stage 2	Quant. of fast initiator (m1) [kg]	Quant. of slow initiator (m2) [kg]	K_w	Porosity	Reaction time (t_r) [min]	MWD
SQP	40 000	Variable (Fig. 3.8)	1.5230	6.6542	70.51	1.91	529.30	57 198
GA	16 000	Variable (Fig. 3.8)	1.5230	6.6542	70.50	1.91	513.70	57 108

The SQP algorithm could barely cope with the dimensions of the problem. After 40 000 function evaluations, the reaction time was reduced by 7 min compared to the best result obtained previously, while the GA after 50 generations (16 000 function evaluations) reduced the reaction time by an additional 23 min (4.3 %) and the MWD by 0.18 %.

By combining results obtained with the optimal fabrication recipe and the optimal temperature profile, total reductions were obtained of 53.8 min (9.5 %) in reaction time, and of 0.4 % in MWD [30].

3.1.4
NMPC of the PVC Batch Reactor

The PVC batch polymerization reactor presents a series of characteristics that make control of the system very difficult:

- The process is discontinuous; hence its physico-chemical properties are changing during the batch, with important modifications of heat and mass transfer. Thus, temperature control becomes difficult.
- The disturbances that occur during the batch, especially when the heat of reaction peaks towards the end of the polymerization, may present challenging control problems, including nonlinear dynamic behavior. At maximum reaction rate, heat removal is very difficult due to increased viscosity of the reaction mixture and polymer deposition on the reactor walls. In this situation, especially if there are other additional disturbances (e.g., variable temperature of the cooling medium), the system may become unstable.
- The system is highly nonlinear (as shown in Fig. 3.9).
- The polymerization being a batch process, there is no steady-state operating point. For this reason, the step response of the system depends on the moment at which the identification of the uncontrolled process is initiated. In order to identify the process needed in the control study, a step input Δu at different moments $t\Delta_u$ was applied, with the restriction that $\Delta u = 0$ for $t < t\Delta_u$. In Figure 3.10, the step response of the system can be seen to vary with $t\Delta_u$, the system being a "time-varying" in nature.

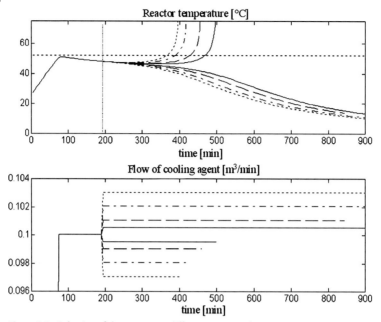

Figure 3.9 Behavior of the system at different steps at the same moment.

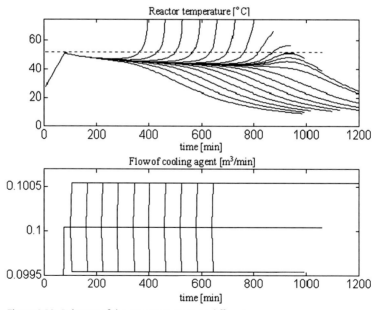

Figure 3.10 Behavior of the system at steps at different moments.

- The behavior of the system depends on the sign of the input step. At a negative step input, the system becomes exponentially unstable, whereas at a positive step input – applied at the same time point – the system behaves in a stable manner. Hence, the system will behave asymmetrically, depending on the control variable.
- The constraint imposed is a variation of temperature of maximum ±0.5 °C around the setpoint in order to ensure proper quality of the product.

All of these process characteristics suggest the use of an advanced nonlinear control method. Several approaches have been proposed to use nonlinear model-based control for batch polymerization systems [2,31,32], and several have been validated on industrial reactors [33]. Here, we present details of two NMPC methods developed for the PVC batch suspension polymerization reactor. A detailed assessment of each approach is presented, together with comparisons of their performance with classical proportional-integral-derivative (PID) control used in practice in the studied plant.

In contrast, it is also shown here that the results of productivity optimization (polymerization at a variable temperature profile) cannot be implemented in practice with conventional PID controllers. However, if an appropriate advanced control strategy is used, the practical implementation of these results is feasible [34].

Valappil and Georgakis (2003) and Nagy and Braatz (2003) each elaborated algorithms to account for batch reactor uncertainty in the nonlinear MPC of end-use properties [35,36]. Likewise, Özcan, either working with Hapoglu and Alpbaz (1998), or in conjunction with Kothare and Georgakis (2003), developed optimal temperature profiles and elaborated a method for the strict control for co-polymerization reactors [37,38]. The problems encountered in the PVC process – namely molar mass and particle size distribution – were also mentioned by Alhamad et al. [39] and by Shi et al., although the latter group were referring to a crystallization process [40].

Here, two NMPC algorithms belonging to the two different classes of methods (see Chapter 2) are presented. The first approach is a modified version of the methods introduced in 1984 by Garcia [41] and in 1992 by Gattu and Zafiriou [42], and belongs to the category of on-line linearization-based control techniques. This approach considers the linearization of the model close to the operating point, the linearized model being used either to solve the optimization problem, or to obtain the "step" or "impulse response", which serves same purpose. The method was applied to the temperature control of a batch polymerization of vinyl chloride monomer (VCM) and tested for different disturbing effects. The results were compared with a second NMPC technique; this was a first-principle model-based NMPC (see Chapter 2, Section 2.3.3) with the sequential solution approach for on-line optimization (see Chapter 2, Section 2.3.6.1). In both NMPC methods the simplified analytical model (presented in Section 3.1.2.3) is used, albeit in two different ways. The first method uses the analytical model to obtain the step response matrix, which is then used to solve the QP problem. The second method uses the first principle nonlinear model directly to solve the QP problem.

3.1.4.1 Multiple On-Line Linearization-Based NMPC of the PVC Batch Reactor

Because the system studied is time-varying in nature, the step response matrix is different at each sampling point and, consequently, it must be regenerated at each sampling time. It also depends on the sign of the step input used to estimate the matrix. Thus, in optimization, one must use the step response matrix obtained for the step with the same sign as the control movement. The proper sign allocation is completed by solving the QP problem twice at each sampling time. First, a prediction is made with $\Delta u(k) = 0$ and $u(k) = u(k + 1) = ... = u(k + P - 1) = u(k - 1)$ for the output of the system at $k - 1$, the corresponding step response matrix being S_{k-1}^k (*model 1*). With S_{k-1}^k, the optimization is solved for the first time, and $\Delta u(k)$ is obtained. If the sampling period is correctly chosen, this $\Delta u(k)$ will be close to – and have the same sign as – the $\Delta u(k)$ obtained from the step response at the moment k. The computed $\Delta u(k)$ $(\Delta u(k)^{(1)})$ is then used to obtain the step response matrix from the rigorous model. This second matrix, which represents the controlled output from the moment k, is used to solve the QP problem for the second time. The result of the optimization, $\Delta u(k)^{(2)}$ is now applied to the process.

The main basis of this method is presented in Figure 3.11 [34,43]. The algorithm is as follows:

Step 1: Measurements at moment k are obtained $(y_m(k))$. We considered in this case all state variables measured or calculated from the measured variables. Thus, the vector $x_m(k)$ of the state measured variables is obtained.

Step 2: Using the rigorous model $\dot{x} = f(x, u); g(x, u) = 0$, a prediction is made by the numerical integration of the model, having the initial conditions $x_m(k)$, $\Delta u(k) = 0$ and $u(k) = u(k + 1) = u(k + P - 1) = u(k - 1)$ Thus, the predicted output at $k - 1$, with $\Delta u(k - 1)$, is obtained:

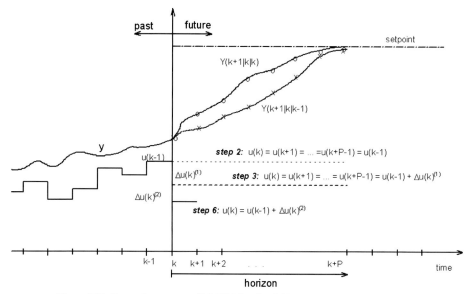

Figure 3.11 The main concept of NMPC1 for the PVC system.

$$Y(k+1|k|k-1) = \begin{bmatrix} y(k+1|k|k-1) \\ y(k+2|k|k-1) \\ \vdots \\ y(k+P|k|k-1) \end{bmatrix} \quad \begin{matrix} u(k) = u(k+1) = \ldots = u(k+P-1) = u(k-1) \\ \Delta u(k) = \Delta u(k+1) = \ldots \Delta u(k+P-1) = 0 \end{matrix}$$

$$(3.153)$$

where $y(/k|k-1)$ are the predictions based on the measurements at moment k and with the step input at $k-1$.

From $Y(k+1|k|k-1)$ the step response matrix which is in fact that one at moment k, but with $\Delta u(k-1)$ is obtained:

$$S_{k-1}^{k} = \begin{bmatrix} \dfrac{y(k+1|k|k-1) - y_m(k)}{\Delta u(k-1)} \\ \dfrac{y(k+2|k|k-1) - y_m(k)}{\Delta u(k-1)} \\ \vdots \\ \dfrac{y(k+P|k|k-1) - y_m(k)}{\Delta u(k-1)} \end{bmatrix}$$

$$(3.154)$$

Step 3: S_{k-1}^{k} (*model 1*) is introduced in an LMPC technique (for example, the *cmpc* function from MathWorks's MPC Toolbox), which solves the QP problem, obtaining:

$$\Delta U(k)^{(1)} = \begin{bmatrix} \Delta u(k)^{(1)} \\ \Delta u(k+1)^{(1)} \\ \vdots \\ \Delta u(k+M-1)^{(1)} \end{bmatrix}$$

$$(3.155)$$

Step 4: A second prediction is made, integrating the model $x = f(x, u)$, with the initial conditions $x_m(k)$, but applying now the step input computed at Step 3. It is obtained thus:

$$Y(k+1|k|k) = \begin{bmatrix} y(k+1|k|k) \\ y(k+2|k|k) \\ \vdots \\ y(k+P|k|k) \end{bmatrix} \quad \begin{matrix} u(k) = u(k+1) = \ldots = u(k+P-1) = u(k-1) + \Delta u(k)^{(1)} \\ \Delta u(k) = \Delta u(k)^{(1)} \\ \Delta u(k+1) = \ldots = \Delta u(k+P-1) = 0 \end{matrix}$$

$$(3.156)$$

In this way, the step response matrix at moment k is obtained:

$$S_{k}^{k} = \begin{bmatrix} \dfrac{y(k+1|k|k) - y_m(k)}{\Delta u(k)^{(1)}} \\ \dfrac{y(k+2|k|k) - y_m(k)}{\Delta u(k)^{(1)}} \\ \vdots \\ \dfrac{y(k+P|k|k) - y_m(k)}{\Delta u(k)^{(1)}} \end{bmatrix}$$

$$(3.157)$$

Step 5: S_k^k (*model 2*) is reintroduced in the LMPC function, which produces the new set of control movements:

$$\Delta U(k)^{(2)} = \begin{bmatrix} \Delta u(k)^{(2)} \\ \Delta u(k+1)^{(2)} \\ \vdots \\ \Delta u(k+M-1)^{(2)} \end{bmatrix} \qquad (3.158)$$

Step 6: The first value from the previously computed vector is sent to the process. Thus, the final control action at moment k is:

$$u(k) = u(k-1) + \Delta u(k)^{(2)} \qquad (3.159)$$

k is incremented to $k+1$ and the above steps are repeated at each sampling time.

For large sampling periods (T) or for regions where the nonlinearity of the process is strong, S_{k-1}^k may be very different to S_k^k; it is possible, after Step 3, to obtain $\Delta u(k)^{(1)}$ with the wrong sign. In such cases, Steps 3, 4, and 5 should be repeated until $sgn(\Delta u(k)^{(i)}) = sgn(\Delta u(k)^{(i-1)})$.

Figure 3.12 illustrates the results of the simulations comparing two ways of solving the QP problem: only once at each sampling point (S_{k-1}^k is used) and twice at each sampling moment (S_{k-1}^k and S_k^k are used). It can be observed that a double prediction is needed.

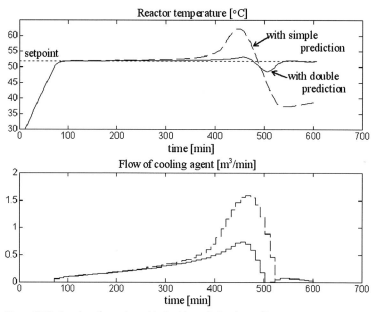

Figure 3.12 Results of running with double and simple prediction.

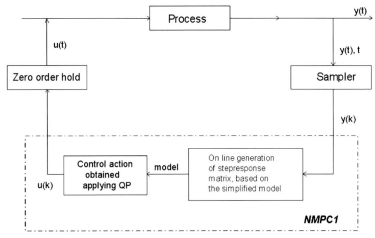

Figure 3.13 NMPC1 block diagram.

Figure 3.13 shows the block diagram of this on-line linearization-based NMPC technique (NMPC1). The flowchart of the implementation of the NMPC1 function in MATLAB programming language is shown in Figure 3.14.

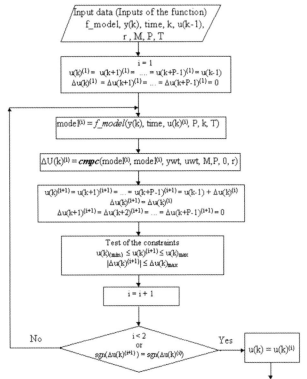

Figure 3.14 Flowchart of NMPC1.

The behavior of the control was studied by simulation, and the results are presented in the next section. In general, this algorithm was seen to provide good results in the first part of the reaction, where the rate is not so high, but worse results in the region of strongest nonlinearity, at maximum reaction rate. In considering the total batch time, the results of this NMPC algorithm can be appreciated as being better than those of the PID system. The proposed method is a modified version of Garcia's QDMC [41], with the main differences being as follows:

- The step response matrix is obtained directly from the rigorous nonlinear model, without any previous linearization.
- At each sampling point, a preliminary detection of the step input sign used to obtain the step response matrix is carried out. This detection is obtained by a double or multiple (if necessary) solving of the QP problem. This feature of the method is advantageous for processes with different types of behavior, depending on the sign of the applied step input.
- The main disadvantage of this method compared to that of Garcia (QDMC) or Gattu and Zafiriou (NLQDMC) [42] is that our NMPC needs all states to be measured or able to be computed from on-line measured variables. However, a direct extension of our method to include a state estimator, is straightforward.
- Another disadvantage is that the QP problem must be solved at least twice at each sampling time, whereas the above-mentioned methods require such solution only once in a sampling period.

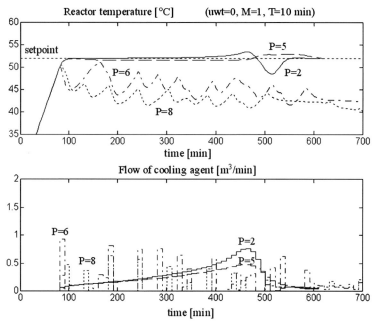

Figure 3.15 Effect of horizon length (P).

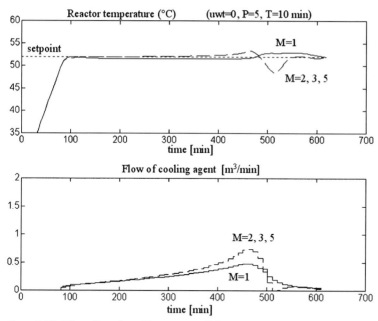

Figure 3.16 Effect of number of input moves (M).

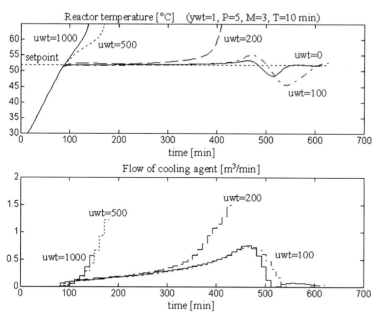

Figure 3.17 Effect of control movement weight (uwt).

The quality of the MPC depends heavily on the parameters of the algorithm: prediction horizon (P); sampling period (T); weight coefficients (uwt, ywt); and number of future moves (*M*). The effects of these parameters are known, but no general tuning methods are available, although some attempts were made [44] (see Chapter 2, Section 2.5 for some general guidelines).

Figures 3.15 to 3.18 illustrate the simulations performed for tuning the MPC parameters for the process considered. It can be observed that:

- The quality of control decreases if *P* exceeds 5 (Fig. 3.15). For *P* = 5 with *M* = 1 and *T* = 10 min, the results are good.
- The value of *M* does not greatly influence the quality of control (Fig. 3.16). The best results were obtained for *M* = 1, but with *M* > 1, the results are comparable. Thus, from the point of view of computational volume, there is no need to choose *M* > 1.
- Figure 3.17 illustrates the fact the value of the *uwt:ywt* ratio strongly influences the quality of control: control is better for lower values of the ratio, the best quality being obtained for *uwt* = 0.
- The effect of the value of sampling period (*T*) is illustrated in Figure 3.18. A very small *T* produces oscillations, while a too-large *T* leads to important control errors, because the step response matrix does not contain sufficient data describing the dynamics of the process. In the case of batch polymerization of VC, a value of *T* = 5–10 min yields the best results.
- For different simulations, the best control results were obtained with different MPC parameters corresponding to the two regions of rate of reaction (lower rate of reaction and highest reaction rate). Based on this observation, an improved MPC was tested with variable *P* and *T* values along the batch. Consequently,

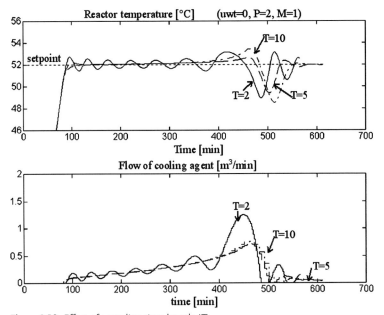

Figure 3.18 Effect of sampling time length (*T*).

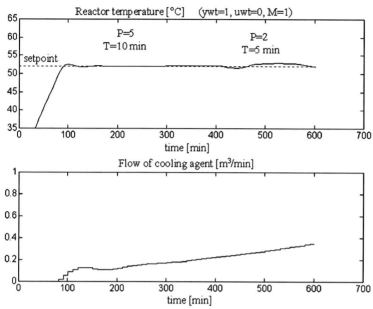

Figure 3.19 The results obtained with variable P and T.

values of $T = 10$ min and $P = 5$ were used until the reaction rate exceeded a certain limit, above which $T = 5$ min and $P = 2$ are used. The data in Figure 3.19 indicate that a very good control performance was achieved.

3.1.4.2 Sequential NMPC of the PVC Batch Reactor

Bequette [45] has pointed out that in future investigations into NMPC the issue of studies comparing two or more NMPC methods at the same process will be of major importance. In this approach, comparison is made between the above-described method and another NMPC using the nonlinear analytical model rather than the linear approach expressed by the step response matrix. Since the process model was highly complex, a sequential method was chosen as this is more easily implemented and requires less computing effort than the simultaneous method. The block diagram of this NMPC (NMPC2) is shown in Figure 3.20.

The *Scope function* computes the optimization objective function, which is the sum of the squares of the differences between the predicted outputs and the setpoint values over the prediction horizon of P time steps for each set of $\Delta U(k)$ obtained from the *optimization algorithm*. In this case, the predicted values $y_p(k + 1)$, are inferred from the numerical integration of the rigorous model. The nonlinear optimization method produces the vector $U(k)$ which satisfies the optimization criteria. Its first value, $u(k)$, is then applied to the input of the process. The golden section method was used as the optimization algorithm, and the control horizon was $M = 1$. For $M > 1$ or for MIMO systems, it is possible to use

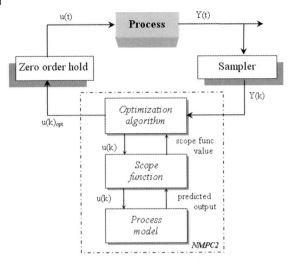

Figure 3.20 Block diagram of the NMPC2 algorithm.

other optimization algorithms (SQP, flexible polyhedron, pattern-search, gradient, etc.). The data in Figure 3.21 indicate that this method is superior to those used previously. A quasi-perfect control can be obtained even in the zone of maximum reaction rate and for a short prediction horizon ($P = 2$, $T_p = 20$ min).

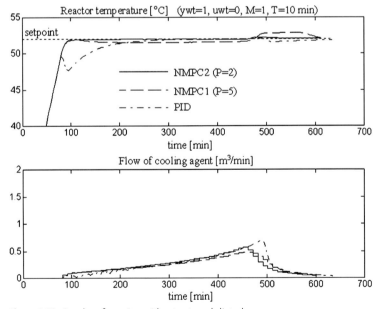

Figure 3.21 Results of running without external disturbances.

The major drawback of this algorithm is that the computation time of the control movement is longer than that with the first method [34]. With the NMPC algorithm it is possible to control the reactor temperature along the variable optimal profile obtained previously, which is impossible with the conventional PID control. In order to obtain the best control performance, the MPC was tuned via simulation by varying the different parameters of the algorithm. Because using a long control horizon significantly increases the computational effort, $M = 1$ was used in all simulations. The best control performance can be obtained with $uwt = 0$, and because the control structure is a SISO one, ywt has no influence on the control performance. Consequently, the parameters that need to be tuned are the sampling time (T_{samp}) and the prediction horizon (P). Figures 3.22 and 3.23 show the effect of these parameters on control performance.

It can observed that the best control performance is achieved with $P = 2$ and $T_{samp} = 3$ min. With these parameters, the optimal temperature profile can be followed almost perfectly, as can be seen in Figure 3.24. This figure also shows that a conventional PID controller is unable to control the process for the presented variable setpoint.

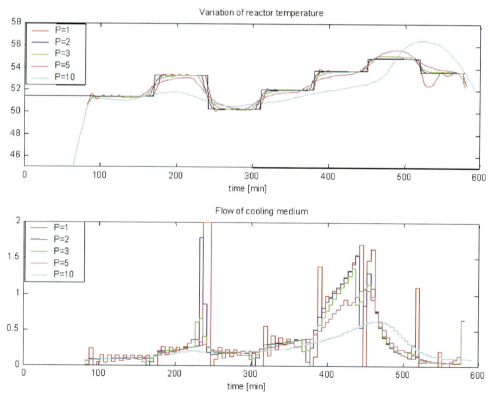

Figure 3.22 Effect of prediction horizon (P) on control performance.

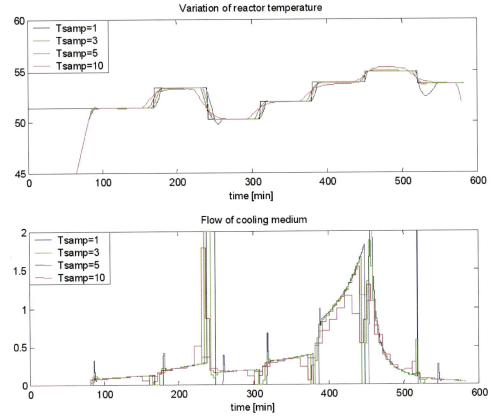

Figure 3.23 Effect of sampling time (T_{samp}) on control performance.

3.1.5
Conclusions

In this section, results concerning the modeling, optimization of productivity and control of the PVC batch reactor have been presented. A rigorous model of this difficult-to-control process was presented, which may be added to a library of dynamic simulations of nonlinear processes. Using this model for optimization, the reaction time and MWD were reduced by obtaining optimal quantities of the slow and fast initiators, as well as by determining an optimal step shape temperature policy. In comparing a SQP and GA to solve optimization problems, the latter was found to outperform the former. By combining results obtained with the optimal fabrication recipe with the optimal temperature profile, it was possible to obtain a total reduction in reaction time of 53.8 min (9.5 %), and in MWD of 0.4 %. Taking into account the number of reactors and extent of production of PVC at the industrial level, these results should, if implemented in practice, lead to significant

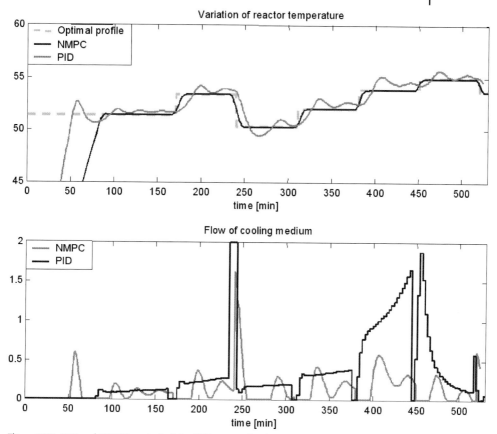

Figure 3.24 PID and NMPC control of the PVC reactor for the optimal temperature profile.

improvements in productivity for an industrial PVC plant. With this in mind, the implementation of advanced NMPC strategies was also studied, and the performance of two nonlinear model predictive control methods and PID control were compared. The first algorithm, based on multiple (double or multiple) on-line linearization in order to cope with nonlinear time-varying processes (NMPC1), yielded good results for the whole batch with the exception of the peak of reaction, where nonlinearity of the system was very strong. The main advantage of this method (NMPC1) was that it allowed the use of MPC linear software. The second NMPC method (NMPC2) used the rigorous model, with the control problem being solved by the sequential method; this yielded, in each case, superior results compared to the NMPC1 and PID control. With this algorithm, even the operation of a process with a frequently changing temperature profile was possible, allowing the PVC reactor to be controlled along its optimal temperature profile. The main disadvantage of the algorithm was the longer time required to obtain control movement. However, with appropriate control hardware, real-time feasibility can be achieved and implemented, and significant increases in reactor productivity will be possible. The software for both methods is written in MATLAB, and is available at the University "Babeş-Bolyai" of Cluj, Romania, Faculty of Chemistry and Chemical Engineering.

3.1.6
Nomenclature

A	weight fraction of monomer contained in polymer
A_{mt}	mass transfer area of monomer from vapors to liquid phase (5.7 m²)
A_T	heat transfer area (20 m²)
c_p	specific heat (J kg⁻¹ K)
D	recycled mass flow of cooling water (2.6 kg s⁻¹)
d_a	diameter of the stirrer (3.4 m)
d_e	equivalent diameter of the jacket (0.5 m)
d_R	diameter of the reactor (2.7 m)
F_{st}	mass flow of the steam in the jacket (kg s⁻¹)
F_w	mass flow of fresh cooling water (kg s⁻¹)
G_{ag}	mass flow of the medium in the jacket (kg s⁻¹)
i_{st}	enthalpy of steam at 6 ata (2.7 × 10⁶ cal kg⁻¹)
k_g	mass transfer coefficient of VC from vapor phase (0.0047 mol m⁻² s·atm)
k_{I1}	fast initiator rate constant (min⁻¹)
k_{I2}	slow initiator rate constant (min⁻¹)
K_{sol}	constant of solubility of monomer in water
K_T	heat transfer coefficient at the wall (W m⁻² K)
k_p	propagation rate constant (m³ kmol⁻¹ min)
k_{tL}	termination rate constant in the liquid phase (m³ kmol⁻¹ min)
k_{tP}	termination rate constant in polymer phase (m³ kmol⁻¹ min)
M	number of future input moves
M_{VC}	monomer molar weight (62.5 kg kmol⁻¹)
MWD	molecular weight distribution
m^0	initial VC quantity (6593 kg)
m_r	mass of reaction (kg)
m_w	mass of water (kg)
n_{I1}	number of kmol of fast initiator (kmol)
n_{I2}	number of kmol of slow initiator (kmol)
n_{rot}	rotation speed of the stirrer (3.35 rot s⁻¹)
P	prediction horizon
Pr	Prandtl criteria number
Pr_p	Prandtl criteria number for the reactor wall
p^0_{VC}	vapor pressure of monomer (atm)
p^e_{VC}	equilibrium pressure of the monomer-water system [atm]
Q	ratio of concentrations of radicals in monomer and polymer (200)
R	universal gas constant (8314 J kg⁻¹ K)
$R(k + 1)$	setpoint matrix

r	setpoint vector
r_{cr}	thermal resistance of the polymer deposit on the inner wall of the reactor ($m^2 K W^{-1}$)
r_T	total thermal resistance of the reactor wall ($m^2 K W^{-1}$)
r_{wall}	thermal resistance of the wall ($5.3 \times 10^{-4} m^2 K W^{-1}$)
S	step response matrix
t	time
t_r	batch time
T, T_{samp}	sampling time
T_{ag}	temperature in the jacket (K)
$T_{in,ag}$	input cooling/heating medium temperature (K)
$T_{out,ag}$	output cooling/heating medium temperature (K)
T_r	temperature in the reactor (K)
T_w	temperature of the fresh water (K)
u	manipulated variable
V_0	initial volume of VC (m^3)
V_j	volume of the jacket ($2.7 m^3$)
V_R	volume of the reactor ($20 m^3$)
V_u	volume of the mass of reaction (m^3)
w	molar fraction
x	state variables
x_m	measured state
$Y(\cdot|\cdot|\cdot)$	predicted values
y	output
y_p	predicted output
y_m	measured output

Greek Symbols

$\alpha_{i,e}$	partial heat transfer coefficients at the wall for its both parts, $[W (m^2 K)^{-1}]$
Γ, Λ	diagonal weight matrices
Γ_l, Λ_l	weight coefficients at moment l
ΔH_r	enthalpy of reaction ($J kg^{-1}$)
$\Delta U, \Delta u$	manipulated variable change
δ_{cr}	thickness of the polymer crust (m)
η	viscosity (Pa·s)
λ	thermal conductivity ($W m \cdot K^{-1}$)
ξ	conversion in the reactor
ξ_c	critical conversion (when the liquid phase of monomer is zero)
ϱ	density ($kg m^{-3}$)
φ	volume fraction of polymer in the gel phase
χ	interaction parameter from Flory–Huggins equation

Subscripts

ag	cooling/heating agent
PVC	polyvinylchloride (polymer)

r	mass of reaction
VC	vinyl chloride (monomer)
w	water

Superscripts

(i)	i^{th} iteration in NMPC1 algorithm
k	k^{th} sampling period
T	transpose

3.2
Modeling, Simulation, and Control of a Yeast Fermentation Bioreactor

3.2.1
First Principle Model of the Continuous Fermentation Bioreactor

Alcoholic fermentation is one of the most important biochemical processes known to man. The attention directed towards this process has increased during the past two decades, mainly because its product – ethanol – might represent an alternative energy source when used as a partial substitute for gasoline as a fuel. Many models of this process have been developed, all of which are based on different kinetic considerations [46–48], although others (e.g., the Maciel Filho group, 2003) have used fuzzy models for nonlinear bio systems identification [49]. Most of these models, however, focus only on the kinetics of the process. The model presented below, when used in simulations beside the detailed kinetic model, involves equations which express the heat transfer, the dependence of kinetic parameters on temperature and the mass transfer of oxygen, as well as the influence of temperature and ionic strength on the mass transfer coefficient. The kinetic equations used in our bioreactor model are modifications of the Monod equations based on Michaelis–Menten kinetics proposed by Aiba and coworkers [50]:

$$\frac{dc_X}{dt} = \mu_X \cdot c_X \cdot \frac{c_S}{K_S + c_S} \cdot e^{-K_P \cdot c_P} \tag{3.160}$$

$$\frac{dc_P}{dt} = \mu_P \cdot c_X \cdot \frac{c_S}{K_{S1} + c_S} \cdot e^{-K_{P1} \cdot c_P} \tag{3.161}$$

$$\frac{dc_S}{dt} = -\frac{1}{R_{SX}} \cdot \frac{dc_X}{dt} - \frac{1}{R_{SP}} \cdot \frac{dc_P}{dt} \tag{3.162}$$

where R_{SX}, R_{SP} are yield factors defined as the ratios of cell and ethanol produced as per the corresponding amount of glucose used for growth or ethanol production, respectively. These equations express the production or consumption of the main components, taking into account the inhibitory effect of ethanol. Our model uses these kinetic equations and describes the continuous fermentation reactor shown schematically in Figure 3.25.

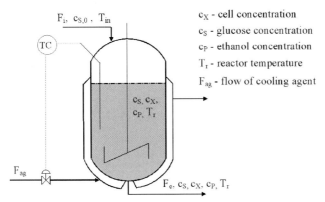

c_X - cell concentration

c_S - glucose concentration

c_P - ethanol concentration

T_r - reactor temperature

F_{ag} - flow of cooling agent

Figure 3.25 The continuous fermentation reactor.

This reactor is modeled as a continuous-stirred tank with constant substrate feed flow. There is also a constant outlet flow from the reactor that contains the product and substrate, as well as the biomass. The reactor contains three distinct main components:

- the biomass, which is a suspension of yeast and which is fed in batch systems and evacuated continuously;
- the substrate, which is a solution of glucose that feeds the micro-organism (*Saccharomyces cerevisiae*); and
- the product, ethanol – this is evacuated together with the other components.

In order to achieve a quasi-steady state with regard to the biomass, a low dilution rate (F_e/V) is necessary – that is, the dilution rate must not exceed the biomass production rate, according to Eq. (3.185). Consequently, the process has a very slow dynamic. Inorganic salts are added together with the yeast, this being necessary for the formation of coenzymes. Due to "salting-out" effects, the inorganic salts also have a strong influence on the equilibrium concentration of oxygen in the liquid phase. This influence of the dissolved inorganic salts, as well as that of temperature, on the equilibrium concentration of oxygen in the liquid phase is modeled in detail by Eqs. (3.163) to (3.180). The mathematical model of the system is presented below.

The initial data of the system are:

- Inorganic salts in the reaction medium:

  $m_{NaCl} = 500g$

  $m_{CaCO_3} = 100g$

  $m_{MgCl_2} = 100g$

- The pH of the liquid phase:

  pH = 6

- The inputs of the system:

$$F_I = Fe = 51 \ L \ h^{-1}$$

$$T_{in} = 25 \ °C$$

$$C_{S,in} = 60 \ g \ L^{-1}$$

$$T_{in,ag} = 15 \ °C$$

- Algebraic equations:
Molar concentrations of ions in the reaction medium are calculated as follows, taking into account that the ion of Cl^- is present in two salts ($NaCl$ and $MgCl_2$):

$$c_{Na} = \frac{m_{NaCl}}{M_{NaCl}} \cdot \frac{M_{Na}}{V} \tag{3.163}$$

$$c_{Ca} = \frac{m_{CaCO_3}}{M_{CaCO_3}} \cdot \frac{M_{Ca}}{V} V\} \tag{3.164}$$

$$c_{Mg} = \frac{m_{MgCl_2}}{M_{MgCl_2}} \cdot \frac{M_{Mg}}{V} \tag{3.165}$$

$$c_{Cl} = \left[\frac{m_{NaCl}}{M_{NaCl}} + 2 \frac{m_{MgCl_2}}{M_{MgCl_2}} \right] \cdot \frac{M_{Cl}}{V} \tag{3.166}$$

$$c_{CO_3} = \frac{m_{CaCO_3}}{M_{CaCO_3}} \cdot \frac{M_{CO_3}}{V} \tag{3.167}$$

$$c_H = 10^{-pH} \tag{3.168}$$

$$c_{OH} = 10^{-(14-pH)} \tag{3.169}$$

The ionic strength of the ion i is calculated using Eq. (3.170):

$$I_i = \frac{1}{2} \cdot c_i \cdot z_i^2 \tag{3.170}$$

$$I_{Na} = 0.5 \cdot c_{Na} \cdot (1)^2 \tag{3.171}$$

$$I_{Ca} = 0.5 \cdot c_{Ca} \cdot (2)^2 \tag{3.172}$$

$$I_{Mg} = 0.5 \cdot c_{Mg} \cdot (2)^2 \tag{3.173}$$

$$I_{Cl} = 0.5 \cdot c_{Ca} \cdot (-1)^2 \tag{3.174}$$

$$I_{CO_3} = 0.5 \cdot c_{CO_3} \cdot (-2)^2 \tag{3.175}$$

$$I_H = 0.5 \cdot c_H \cdot (1)^2 \tag{3.176}$$

$$I_{OH} = 0.5 \cdot c_{OH} \cdot (-1)^2 \tag{3.177}$$

The global effect of the ionic strengths is given by Eq. (3.178):

$$\sum H_i \cdot I_i = H_{Na} \cdot I_{Na} + H_{Ca} \cdot I_{Ca} + H_{Mg} \cdot I_{Mg} + H_{Cl} \cdot I_{Cl} + \dots$$
$$+ H_{CO_3} \cdot I_{CO_3} + H_H \cdot I_H + H_{OH} \cdot I_{OH} \qquad (3.178)$$

The dependence of the equilibrium concentration of oxygen with temperature in distilled water is given by the empirical equation below, which is obtained from the experimental data presented by Sevella [51]:

$$c_{O_2,0}^* = 14.6 - 0.3943 \cdot T_r + 0.007714 \cdot T_r^2 - 0.0000646 \cdot T_r^3 \qquad (3.179)$$

Due to the fact that salts are dissolved in the medium, the equilibrium concentration of oxygen in liquid phase is obtained from the following Setchenov-type equation:

$$c_{O_2}^* = c_{O_2,0}^* \cdot 10^{-\sum H_i \cdot I_i} \qquad (3.180)$$

The mass transfer coefficient for oxygen, and the temperature function, is given by the following empirical equation (Sevella, 1992):

$$(k_l a) = (k_l a)_0 \cdot (1.204)^{T_r - 20} \qquad (3.181)$$

The rate of oxygen consumption is:

$$r_{O_2} = \mu_{O_2} \cdot \frac{1}{Y_{O_2}} \cdot c_X \cdot \frac{c_{O_2}}{K_{O_2} + c_{O_2}} \qquad (3.182)$$

Expression of the maximum specific growth rate [Eq. (3.183)] involves the resultant of the growth rate increasing with the temperature and the effect of the heat denaturation:

$$\mu_X = A_1 \cdot e^{-\frac{E_{a1}}{R(T_r + 273)}} - A_2 \cdot e^{-\frac{E_{a2}}{R(T_r + 273)}} \qquad (3.183)$$

- Differential equations:
The balance for the total volume of the reaction medium is:

$$\frac{dV}{dt} = F_i - F_e \qquad (3.184)$$

The mass balances for the biomass, product, substrate and dissolved oxygen are expressed by Eqs. (3.185) to (3.188):

$$\frac{dc_X}{dt} = \mu_X \cdot c_X \cdot \frac{c_S}{K_S + c_S} \cdot e^{-K_P \cdot c_P} - \frac{F_e}{V} c_X \qquad (3.185)$$

$$\frac{dc_P}{dt} = \mu_P \cdot c_X \cdot \frac{c_S}{K_{S1} + c_S} \cdot e^{-K_{P1} \cdot c_P} - \frac{F_e}{V} c_P \qquad (3.186)$$

The first terms in Eqs. (3.185) and (3.186) represent the quantity of biomass and product, respectively, produced in the fermentation reactions. The last terms describe the amount of yeast and ethanol respectively, leaving the reactor.

$$
\frac{dc_S}{dt} = \frac{1}{R_{SX}} \cdot \mu_X \cdot c_X \cdot \frac{c_S}{K_S + c_S} \cdot e^{-K_P \cdot c_P} -
$$
$$
- \frac{1}{R_{SP}} \cdot \mu_P \cdot c_X \cdot \frac{c_S}{K_{S1} + c_S} \cdot e^{-K_{P1} \cdot c_P} + \frac{F_i}{V} c_{S,in} - \frac{F_e}{V} c_S \tag{3.187}
$$

The first and second terms in Eq. (3.187) represent the amount of substrate consumed by the biomass for growth and ethanol production, respectively. The third term is the quantity of glucose entering the reactor with the fresh substrate feed, while the last term is the quantity of glucose leaving the reactor.

The concentration of the dissolved oxygen in reaction medium is the resultant of the quantity of oxygen entering in the reaction medium due to the mass transfer, expressed by the first term in Eq. (3.188), and the amount consumed in the fermentation reactions (last term):

$$
\frac{dc_{O_2}}{dt} = (k_l a) \cdot \left(c_{O_2}^* - c_{O_2} \right) - r_{O_2} - \frac{F_e}{V} \cdot c_{O_2} \tag{3.188}
$$

The energy balances for the reactor and jacket are given by Eqs. (3.189) and (3.190), respectively.

$$
\frac{dT_r}{dt} = \frac{F_i}{V} \cdot (T_{in} + 273) - \frac{F_e}{V} \cdot (T_r + 273) + \frac{r_{O_2} \cdot \Delta H_r}{32 \cdot \varrho_r \cdot C_{heat,r}} +
$$
$$
+ \frac{K_T \cdot A_T \cdot (T_r - T_{ag})}{V \cdot \varrho_r \cdot C_{heat,r}} \tag{3.189}
$$

$$
\frac{dT_{ag}}{dt} = \frac{F_{ag}}{V_j} \cdot (T_{in,ag} - T_{ag}) + \frac{K_T \cdot A_T \cdot (T_r - T_{ag})}{V_j \cdot \varrho_{ag} \cdot C_{heat,ag}} \tag{3.190}
$$

The parameters of the model are presented in Table 3.5 [52].

Table 3.5 Parameters of the process model.

$A_1 = 9.5 \cdot 10^8$	$H_{Cl} = 0.844$	$R_{SP} = 0.435$
$A_2 = 2.55 \cdot 10^{33}$	$H_{CO_3} = 0.485$	$R_{SX} = 0.607$
$A_T = 1 \, m^2$	$H_{HO} = 0.941$	$V = 1000 \, L$
$C_{heat,ag} = 4.18 \, J \cdot g^{-1} K^{-1}$	$(k_l a)_0 = 38 \, h^{-1}$	$V_j = 50 \, L$
$C_{heat,r} = 4.18 \, J \cdot g^{-1} K^{-1}$	$K_{O_2} = 8.86 \, mg \, L^{-1}$	$Y_{O_2} = 0.970 \, mg \, mg^{-1}$
$E_{a1} = 55\,000 \, J \, mol^{-1}$	$K_P = 0.139 \, g \, L^{-1}$	$\Delta H_r = 518 \, kJ \, mol^{-1} \, O_2$
$E_{a2} = 220\,000 \, J \, mol^{-1}$	$K_{P1} = 0.070 \, g \, L^{-1}$	$\mu_{O_2} = 0.5 \, h^{-1}$
$H_{Na} = -0.550$	$K_S = 1.030 \, g \, L^{-1}$	$\mu_P = 1.790 \, h^{-1}$
$H_{Ca} = -0.303$	$K_{S1} = 1.680 \, g \, L^{-1}$	$\varrho_{ag} = 1000 \, g \, L^{-1}$
$H_{Mg} = -0.314$	$K_T = 3.6 \cdot 10^5 \, J \cdot h^{-1} m^{-2} K^{-1}$	$\varrho_r = 1080 \, g \, L^{-1}$
$H_H = -0.774$		

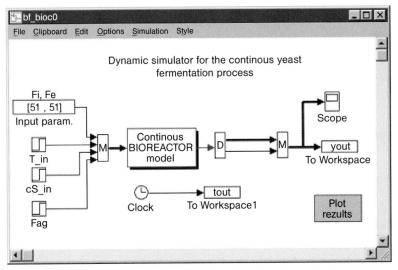

Figure 3.26 SIMULINK block diagram of the model.

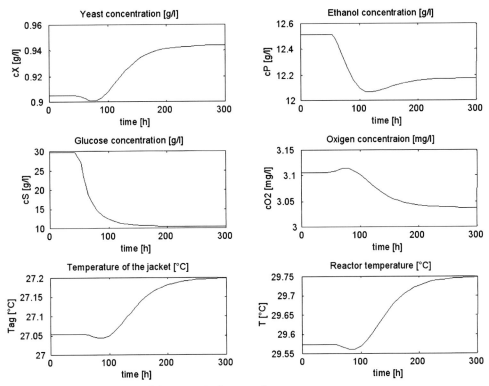

Figure 3.27 Dynamic response of the system in the case of step change in the input substrate concentration ($40 \rightarrow 60$ g L^{-1}).

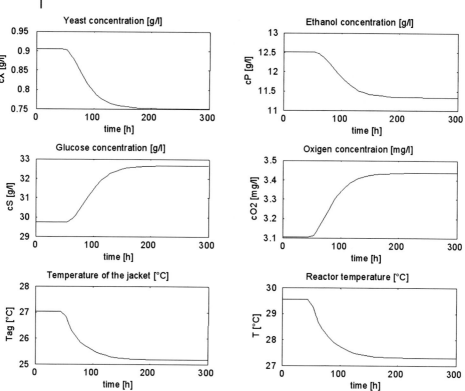

Figure 3.28 Dynamic response of the system in the case of step change in the temperature of input flow (25→23 °C).

The above-described model was used for studying the dynamic behavior of the reactor in the case of different disturbances. The disturbances considered were: step change in the inlet flow temperature and in the substrate concentration. The first disturbance can occur due to the ambient temperature variation, while the second can occur as a result of quality changes of the substrate flow.

The model was implemented as a MATLAB S-function. The SIMULNK block diagram is represented in Figure 3.26.

The dynamic behavior of the system for the disturbances studied is shown in Figures 3.27 and 3.28.

It can be observed that an important variation in the input concentration causes a slight variation in the ethanol concentration, so that this disturbance will not be considered as major.

According to the data in Figure 3.28, the effect of the change in the inlet temperature is of much greater importance, with a step of only 2 °C causing important variations in ethanol concentrations.

The presented model can serve as a valuable tool for testing a variety of control methods. Conventional PID control, LMPC and artificial neural network (ANN) model-based NMPC (AANMPC) control techniques are tested in the following section.

3.2.2
Linear Model Identification and LMPC of the Bioreactor

The golden rule in identification is to try simple things first. If a linear model is capable of describing the dynamic behavior of the system, there is no need to examine the use of nonlinear models. In order to determine how a linear model describes the process, two such models were identified based on simulated data obtained from the analytical model described in Section 3.2.1:
- The *linear state space model*, expressed by the equations below:

$$X(k+1) = \Phi \cdot X(k) + \Gamma \cdot U(k)$$
$$Y(k) = C \cdot X(k) \tag{3.191}$$

  where Φ, Γ and C are discrete state space matrices for the corresponding sampling time, $X(k)$ is the state vector, $Y(k)$ the output vector of the linear model, and $U(k)$ is the vector of the manipulated variables at moment k. With this model the process nonlinearity is demonstrated by the simulation results presented in Figure 3.29. The procedure of obtaining the plotted data in Figure 3.29 was as follows. At the steady state operating point (where $F_{ag} = 18$ L h^{-1} and $T_r \cong 30$ °C), a sequence of step inputs (ΔF_{ag}) was given. The changes in output (ΔT_r) after one sampling period are depicted, and the difference between the nonlinear model (represented by circles) and the linearized model (solid line) is clear.
- The *Output-Error (OE)* model, from the polynomial linear model category was also identified using the simulated input-output data pairs:

$$y(k) = -f_1 \cdot (y(k-1) - e(k-1)) - f_2 \cdot (y(k-2) - e(k-2)) - \dots$$
$$-f_{nf} \cdot (y(k-nf) - e(k-nf)) + \dots + b_1 \cdot u(k-nk) + \tag{3.192}$$
$$b_2 \cdot u(k-nk-1) + \dots + b_{nb} \cdot u(k-nk-nb+1) + e(k)$$

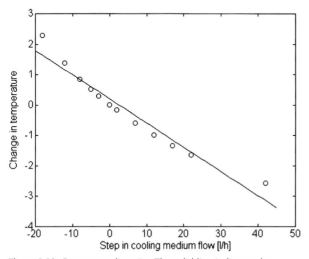

Figure 3.29 Process nonlinearity. The solid line indicates the temperature changes for a linear approximation of the process.

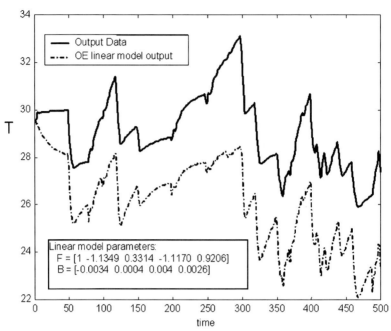

Figure 3.30 Output error (OE) linear model prediction.

where $f_1, f_2, ..., f_{nf}, b_1, b_2, ..., b_{nb}$, are the coefficients of the model. The structure of the model is defined by giving the time delay nk, and the order of the polynomials nf and nb, respectively. In order to assure a sufficient complexity of the model, a structure with parameters $nk = 1$, $nf = nb = 4$, was identified. Figure 3.30 shows that the obtained linear model is unable to model the process accurately; in spite of that, with feedback model/plant correction even the linear model-based LMPC might be capable of controlling the process. However, because the underlying system is nonlinear, one should expect that the nonlinear model-based controller would give better control performances. Consequently, the development of an accurate nonlinear model is justified.

Although the linear models do not describe very accurately the dynamic behavior of the bioreactor, when the disturbance does not move the process far from the operation point where the linear model was obtained, fairly good control performance can be expected.

The disturbance caused by the input temperature change was used to study the performances of two different controllers. The reactor temperature was the controlled variable, and the flow of cooling medium was the manipulated input. First, an advanced "anti-windup" digital PID controller with position algorithm and backward approximation was used. The tuning parameters were $K = 110$,

Ti = 60, and Td = 0. The very good results obtained with this controller are shown in Figure 3.31 [53].

Despite these good results, an attempt was made to improve the performances by implementing a more advanced control structure – the model predictive control. The linear model was first represented by the step response matrix; the results obtained with this LMPC algorithm are presented in Figure 3.32. The main tuning parameters used for the LMPC controller were $P = 5$; $M = 2$; $\Gamma = [1]$; $\Lambda = [0]$. It can be seen that an excellent control performance was achieved, with both the deviation peak from the setpoint and the settling time being smaller in the case of LMPC than with PID control. However, the simulation results showed that even the advanced PID control gave fairly good results. This can be explained by the fact that the studied disturbances do not push the process far from the operating point, where the assumption of linear behavior is valid. A worse control performance should be expected for a variable setpoint (which forces the process operate in a large range of operating points) or measurement noise. In this case, a nonlinear model based control technique might be necessary.

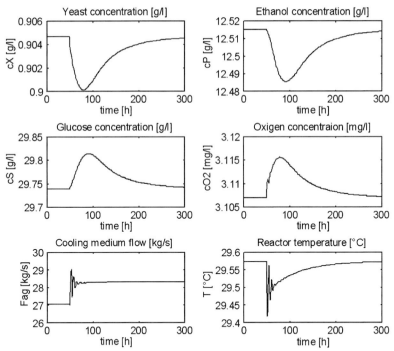

Figure 3.31 Simulation results with PID control of the process.

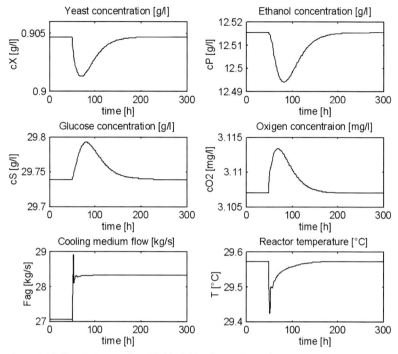

Figure 3.32 Simulation results with Model Predictive Control (MPC) of the process.

3.2.3
Artificial Neural Network (ANN)-Based Dynamic Model and Control of the Bioreactor

3.2.3.1 Identification of the ANN Model of the Bioreactor

Chemical processes in general – and biochemical fermentation systems in particular – have strongly nonlinear features, as was demonstrated previously. In addition, biochemical process models have many parameters that must be determined experimentally [54,55]. For these reasons, the linear model approach is on the one hand inappropriate for such processes, yet on the other hand accurate analytical model development may be very difficult to achieve, and its use in a NMPC scheme requires much computational effort and time [56,57]. Consequently, neural networks represent a valuable tool in the modeling and control of biochemical process.

Here, the primary goal was to obtain a dynamic ANN model which describes the variations of the reactor temperature as a function of the cooling agent flow. For this, a random input signal was generated and applied to the system. The simulated response of the system, together with the random input signal, was used to train the ANN. Once the ANN model is identified, it can be used as an internal model in an advanced nonlinear model predictive control algorithm. For this, it is crucial to have a network with very good generalization properties. One way to obtain a network with appropriate generalization properties is to choose a structure with sufficient parameters; this ensures that it learns the training data and then optimizes the topology of the network until the best generalization properties are achieved.

Consequently, a feed-forward neural network was chosen, with the same *nk, nf* and *nb* parameters as in the case of the linear OE model. Thus, in the input layer the network has eight neurons, while there is one neuron in the output layer, with linear transfer function. One hidden layer with 14 neurons with the hyperbolic tangent sigmoid transfer function was used. The fully connected initial network is presented in Figure 3.33, while the input-output data sequence used to train the network is shown in Figure 3.34.

The network was trained using the Levenberg–Marquardt training algorithm. Figures 3.35 and 3.36 illustrate that the network was able to learn the training data with an exceptional accuracy.

In order to test the ANN capability of generalization, another random input sequence was obtained, by simulation from the first principle model of the system. The test data set is presented in Figure 3.37, while Figures 3.38 and 3.39 demonstrate the very pure generalization performance of the ANN model. In this case, very high prediction errors were obtained. By comparing the plots for training and the test set, it is quite clear that the network is overfitting the data. It is concluded, therefore, that the model structure selected contains too many neurons (weights). Consequently, in order to improve the generalization performance of the ANN model, it is necessary to remove the superfluous weights from the network.

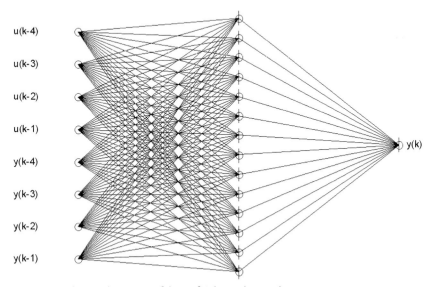

Figure 3.33 The initial structure of the artificial neural network (ANN).

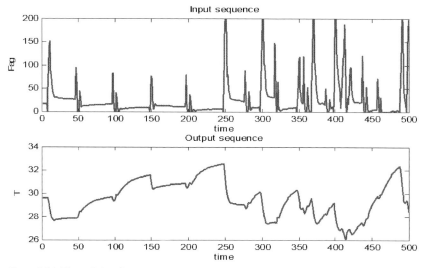

Figure 3.34 The training data.

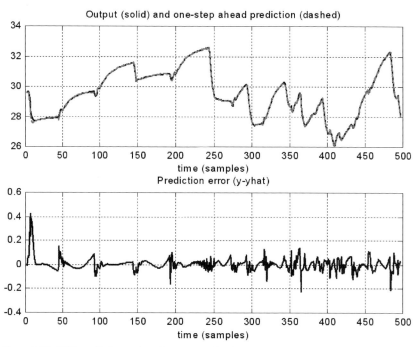

Figure 3.35 ANN prediction and prediction error for the training data.

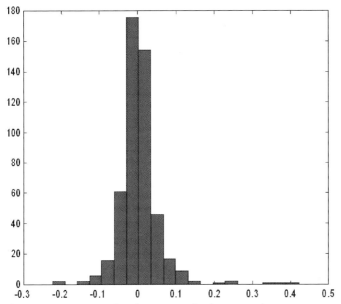

Figure 3.36 Histogram of prediction errors for the training data.

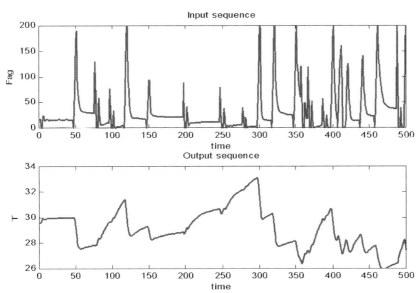

Figure 3.37 Data set to test the ANN.

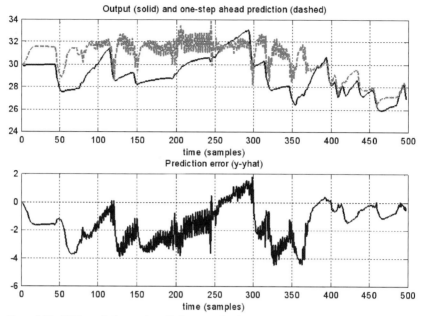

Figure 3.38 ANN prediction and prediction error for the testing data.

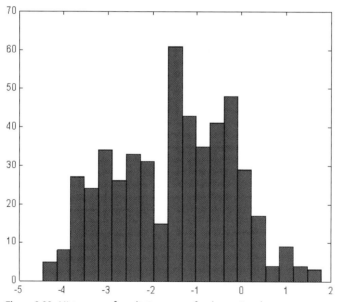

Figure 3.39 Histogram of prediction errors for the testing data.

3.2.3.2 Using Optimal Brain Surgeon to Determine Optimal Topology of the ANN-Based Dynamic Model

3.2.3.2.1 Description of the Enhanced OBS Algorithm

One of the most important parameters of the ANN is the number of connections among the neurons. As can be seen on the simulation results presented in Section 3.2.3.1, this parameter determines the learning and especially the generalization performances of the ANN. Recently, many studies have been conducted to investigate improvements in network training by using different network structures, transfer functions and learning algorithms [58–61], as well as by elucidating definite methodologies to determine the network structures [62].

The so-called Optimal Brain Surgeon (OBS) is the most important strategy for pruning neural networks. This algorithm determines the optimal network architecture by removing any superfluous weights from the network in order to avoid overfitting of the data by the ANN. We implemented the OBS algorithm proposed by Hansen and Pedersen [63], and modified it to take into account that it should not be possible to have networks where a hidden unit has lost all the weights leading to it, while there still are weights connecting it to the output layer, or vice-versa.

In this algorithm a saliency is defined as the estimated increase of the unregularized error criterion, when a weight is eliminated. The saliency for weight j is defined by:

$$\xi_j = \lambda_j I_j^T H^{-1}(\theta^*) \frac{1}{N} Q\theta^* + \frac{1}{2} \lambda_j^2 I_j^T H^{-1}(\theta^*) R(\theta^*) H^{-1}(\theta^*) I_j \tag{3.193}$$

where θ^* is a vector with all the weights and biases of the reduced network and I_j is the jth unit vector. The Gauss–Newton Hessian of the regularized criterion is calculated with Eq. (3.194):

$$H(\theta^*) = R(\theta^*) + \frac{1}{N} Q \tag{3.194}$$

where R is the Hessian of the unregularized error criterion, and Q is the regularization matrix. The Lagrange multiplier λ_j is calculated from the following equation:

$$\lambda_j = \frac{\theta_j^*}{H_{j,j}^{-1}(\theta^*)} \tag{3.195}$$

The constrained minimum (the minimum when weight j is 0) is then found from:

$$\delta\theta = \theta^* - \theta = -\lambda_j H^{-1}(\theta^*) I_j \tag{3.196}$$

Initially, the saliences are calculated and the weights pruned as described above. However, when a situation occurs where a unit has only one weight leading to or one weight leading from it, the saliency for removing the entire unit is calculated instead, by setting all weights connected to the unit to zero. With the proposed enhanced OBS algorithm the computational time necessary to obtain the optimal topology was reduced in some cases by 30%. The network can be retrained after each weight elimination or after a certain percentage of the weights had been eliminated. The error criterion used by the algorithm is calculated for the test data set.

This algorithm was implemented in MATLAB and successfully applied to the above-obtained ANN model.

3.2.3.2.2 Simulation Results of Pruning the ANN Model with the OBS Algorithm

The fully connected feed-forward ANN, as used in our simulations, contains a total number of 141 parameters (weights and biases). The OBS algorithm was used in order to prune the network. After each weight elimination the network was retrained for 50 iterations. Figure 3.40 presents the results obtained with the OBS algorithm. In this figure, the error criteria for both the training data and testing data, together with the final prediction error (FPE), are presented. The FPE is estimated from the training set, and is very useful when a test set is not available. The test error is the most reliable estimate of the generalization error; therefore, the OBS algorithm selects the network with the smallest test error. The OBS algorithm gave as the final network the one with 53 weights (a reduction of 88 weights, i.e., 62%). This result can be confirmed by examining Figure 3.40, while the architecture of the selected network is presented in Figure 3.41.

It can be seen that a considerable reduction in network structure was achieved, with the number of the weights being reduced by 62%, and three neurons from the

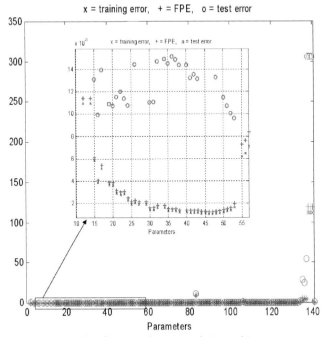

Figure 3.40 Results of pruning the ANN with Optimal Brain Surgeon (OBS).

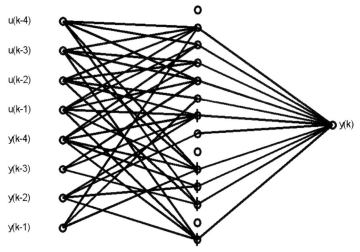

Figure 3.41 The architecture of the optimal ANN (with 53 weights) obtained by the OBS algorithm.

hidden layer being completely eliminated. The performances on the testing data (Figs. 3.42 and 3.43), obtained with the reduced ANN show great promise.

Further study of the results obtained with the OBS algorithm (see Fig. 3.40) shows that the network with the second-best test error has a much simpler architecture (16 weights). The test error is very close to that obtained by the

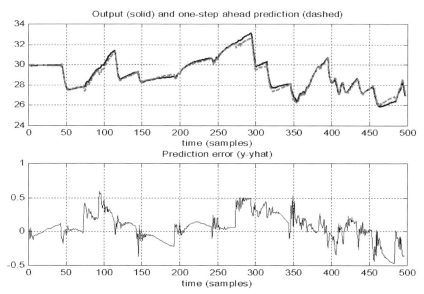

Figure 3.42 Generalization performances of the reduced ANN (with 53 weights).

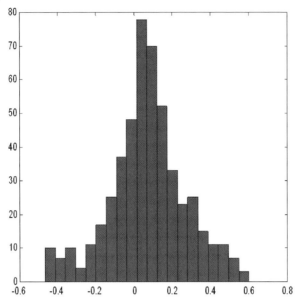

Figure 3.43 Histogram of prediction errors for the testing data
(ANN with 53 weights).

ANN with 53 weights, but the topology is reduced additionally by 37 weights. The
structure of this ANN is presented in Figure 3.44. This network suffered considerable simplification of its structure, with 125 weights being removed from a total
of 141 (a reduction of 88%). Three neurons were eliminated from the input layer,
and seven from the hidden layer. This structure shows that, for accurate modeling

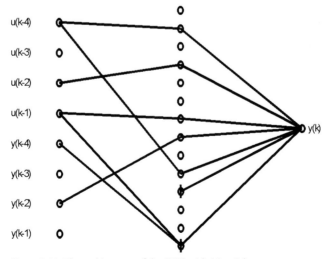

Figure 3.44 The architecture of the ANN with 16 weights.

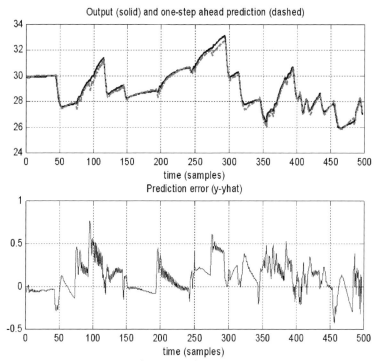

Figure 3.45 Generalization performances of the reduced ANN (with 16 weights).

of the dynamic behavior of the system, there is no need to use either the output measurements with one and three past steps, or the input value with three past sampling times.

Despite the considerable reduction in network structure complexity, this simple structure provides very good generalization properties (see Fig. 3.45) and facilitates rapid data processing. Consequently, the pruned networks may be used in different nonlinear model predictive or optimal control algorithms as the internal model used for prediction.

3.2.3.3 ANN Model-Based Nonlinear Predictive Control (ANMPC) of the Bioreactor

Once the ANN model is identified and the structure with the best generalization properties selected, it can be used in different NMPC algorithms. In NMPC, multi-step-ahead prediction is usually needed to foresee the behavior of the process. For this, it is possible to use the previously identified ANN model with one-step-ahead prediction (in this case multi-step-ahead prediction can be obtained using the ANN repeatedly) or another structure with more than one future output parameter in the output layer. The latter has the advantage of faster computation of predicted values, but in this case the prediction horizon is usually fixed when the network structure is chosen. For different prediction horizons, it will be necessary to train different networks (for details, see Chapter 2, Section 2.3.5).

In this section we exemplify the implementation of ANMPC with a *non-recursive* four step-ahead predictor ANN model. A net with two hidden layers with seven and five neurons respectively, was trained. The network had six input neurons corresponding to the current and past (with two sampling periods) values of the reactor temperature (controlled variable) and flow of cooling agent (manipulated variable). The prediction horizon of the network was equal four sampling periods – that is, it had four neurons in the output layer. This network can be represented abstractly as follows:

$$[T_r(k+1), T_r(k+2), T_r(k+3), T_r(k+4)] = f_{NN}\big(F_{ag}(k), \dots$$
$$F_{ag}(k-1), F_{ag}(k-2), T_r(k), T_r(k-1), T_r(k-2)\big) \tag{3.197}$$

The sigmoid transfer function was used in the hidden layers, as well as in the output layer. The network was trained for 10 000 epochs with the back-propagation learning algorithm, using the historical database of plant inputs and outputs obtained from the analytical model of the process. After learning, the network gave very good prediction and generalization performances.

The network, once defined and trained, was introduced in a model predictive control scheme as the internal model used for prediction during the control movement calculation. The neural network model-based predictive control (ANMPC) structure is presented in Figure 3.46.

In each sampling period the current temperature measurement is obtained $(T_r(k))$ and, considering that the past temperature measurements and control actions are known, the next control action is determined by solving an optimization problem. That is, the next control action is selected such that the predicted outcome of the control action is optimum in the sense of minimizing the square of the

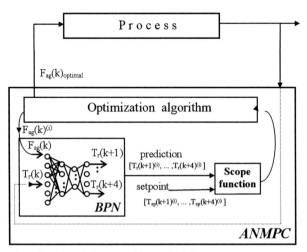

Figure 3.46 Block diagram of ANMPC of the process.

deviation from the setpoint trajectory over a finite horizon (*P*). Consequently, the optimization problem for this particular case can be formulated as follows:

$$\min_{F_{ag}(k)} \left\{ \sum_{i=1}^{P} \left[T_r(k+i) - T_{sp}(k+i) \right]^2 \right\} \tag{3.198}$$

where:

$$T_r(k+i)|_{i=\overline{1,P}} = f_{NN}\left(F_{ag}(k), F_{ag}(k-1), F_{ag}(k-2), T_r(k), T_r(k-1), T_r(k-2)\right) \tag{3.199}$$

With this control structure, an excellent control of the process was achieved. Here, a control horizon *P* = 4 was used. For *P* > 4, the procedure for non-recursive *d*-step ahead prediction described in Chapter 2 must be used. For comparison, the PID and LMPC control of the process are also presented in Figure 3.47, where the superiority of ANMPC can be clearly seen.

The robustness of the ANMPC structure was studied under conditions of noisy temperature measurement. The amplitude of the white noise in the simulation was 1.5 °C. In order to render the network capable of controlling the process in the case of noisy temperature measurement, it was trained by including noisy data into the training set. If I and D are the training input, respectively output, obtained

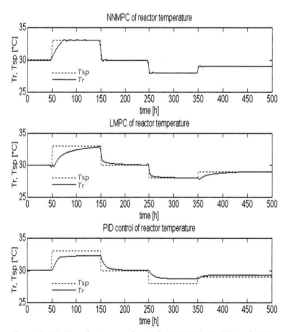

Figure 3.47 Simulation results with ANMPC, LMPC and PID control of the process.

from the analytical model of the process, the training set for training the net with noise can be developed as follows:

– the input data: $\qquad\qquad\qquad\mathbf{I}^* = [\mathbf{PP} + noise]$ (3.200)

– the corresponding target data: $\quad\mathbf{D}^* = [\mathbf{DD}]$ (3.201)

When the net had been trained with these training data it was used in the above-described ANMPC scheme. The results obtained in the case of noisy temperature measurement are shown in Figure 3.48, with a fairly good control being achieved. In this case, PID control gave worse results than ANMPC, while LMPC failed totally for every filtering coefficient utilized.

In this example, as an alternative approach to using neural networks for process control, the use of an inverse neural network model is considered. Here, the outputs of the net correspond to the future values of the process inputs, while the input layer of the net contains, in addition to past values of the process inputs and outputs and the current process output measurement, also the future values of controlled variables (process outputs). Due to its structure, the inverse neural network eliminates the optimization algorithm from the control movement computation. Using past values of controlled and manipulated variables as well as the current measurement, the control movement can be obtained directly from the net when the setpoint values are presented to the network as the future values of the

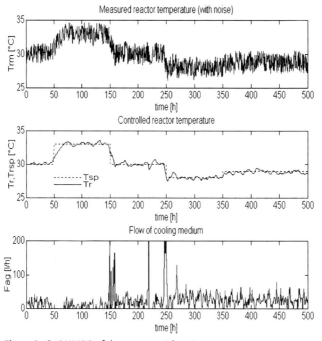

Figure 3.48 ANMPC of the process with noisy temperature measurement.

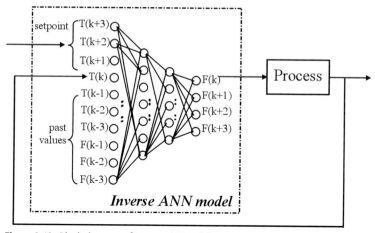

Figure 3.49 Block diagram of Inverse-ANN of the process.

controlled variables. The block diagram of the inverse neural network model-based predictive control (IANMPC) of the process is presented in Figure 3.49.

The network used in these simulations had two hidden layers (with seven and five neurons, respectively), with 10 neurons in the input layer and four in the output layer. It can be represented as:

$$
\begin{aligned}
\left[F_{ag}(k), F_{ag}(k+1), F_{ag}(k+2), F_{ag}(k+3)\right] &= f_{invNN}(T_r(k+3), T_r(k+2), T_r(k+3), \\
&\quad T_r(k+1), \ldots, T_r(k), T_r(k-1), T_r(k-2), \\
&\quad T_r(k-3), F_{ag}(k-1), F_{ag}(k-2), F_{ag}(k-3))
\end{aligned}
$$

$$(3.202)$$

The network was trained using an historical database obtained from the analytical model. In the training phase, the future values of temperature ($T_r(k+i)$) are known. When the network had been trained it was used for control when, at each sampling time, for the future temperature inputs of the network the setpoint values were presented:

$$
T_r(k+1)\big|_{i=\overline{1,3}} = T_{sp}(k+1)\big|_{i=\overline{1,3}}
$$

The data presented in Figure 3.50 show that a very good control performance was achieved with this control structure.

3.2.4
Conclusions

This section has successfully demonstrated the ability of ANNs to model complex nonlinear biochemical processes, such as alcoholic fermentation. The detailed analytical model of the continuous fermentation bioreactor was presented. This

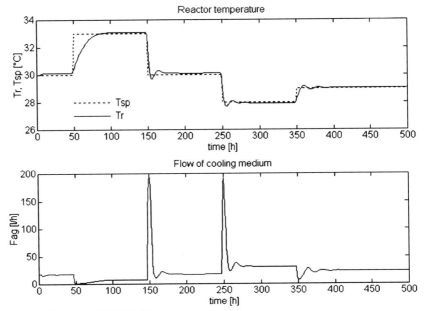

Figure 3.50 Inverse-ANMPC of the process.

model is more complex than those used generally to test different control systems; therefore, it involves more nonlinear characteristics of the process. Consequently, the presented model may be valuable when testing a variety of nonlinear control methods. Using the data obtained from the analytical model, ANN-based models were also developed.

An efficient new algorithm – the enhanced Optimal Brain Surgeon (OBS) – was also presented as a pruning algorithm for the determination of optimal ANN topology. With the OBS algorithm a reduction in the number of weights from 141 to 53 (–62%) in the first step, and finally to 16 (–88%) was achieved. Simulation results were presented to show that this very simple network structure can perform a better generalization than the initial fully connected one. The pruned networks have very good generalization properties [64], with the simple structure facilitating rapid data processing. Consequently, the pruned networks can be used in different nonlinear model predictive algorithms as the internal model used for prediction.

Two ANN model-based NMPC schemes were presented and tested via simulations, and the results compared with those obtained with linear MPC and PID control [65,66]. The superiority of the ANMPC structure was demonstrated. The main advantage of the ANMPC compared to the NMPC (which uses the analytical model of the system) was that it needs minimal knowledge of the process – a feature that might be crucial in the case of biochemical processes. The nonlinear model used in the ANMPC can be obtained from experimental input-output data, and it is much more simple than its analytical counterpart.

It appears that the development of the ANMPC is easier than that of the NMPC, but that because the control movement calculation is iterative it is faster than in the case of NMPC, which requires online integration of the complex analytical model.

The simulations presented also demonstrate how neural networks can be trained and used for nonlinear model predictive control of the process when measurements are affected by noise.

Additionally, in this chapter, the development and application of a predictive control scheme based upon the inverse neural network model of the process was also presented. The main advantage of this control structure is that it requires an extremely simple mathematical apparatus for control movement calculation. As this algorithm is no longer iterative, the required computational time is very short, making it preferable for real-time applications.

3.2.5
Nomenclature

A_j	the output of the jth layer of the network
A_N	the output of the neural network
A_1, A_2	pre-exponential factors in Arrhenius equation
A_T	heat transfer area (m^2)
C	discrete state space matrix used in linear models
$C_{heat,ag}$	heat capacity of cooling agent (J·g^{-1} K^{-1})
$C_{heat,r}$	heat capacity of mass of reaction (J·g^{-1} K^{-1})
c_j	concentration of ion j (j = Na, Ca, Mg, Cl, CO$_3$, etc.)
c_{O_2}	oxygen concentration in the liquid phase (mg L^{-1})
$c_{O_2}^*$	equilibrium concentration of oxygen in the liquid phase (mg L^{-1})
$c_{O_2,0}^*$	equilibrium concentration of oxygen in distilled water (mg L^{-1}l)
c_P	product (ethanol) concentration (g L^{-1})
c_S	substrate (glucose) concentration (g L^{-1})
$c_{S,in}$	glucose concentration in the feed flow (g L^{-1})
c_X	biomass (yeast) concentration (g L^{-1})
D	output data from the training set
E	sum squared error of the network
E_{a1}, E_{a2}	apparent activation energy for the growth respectively denaturation reaction
F_{ag}	flow of cooling agent (L h^{-1})
F_e	outlet flow from the reactor (L h^{-1})
F_i	flow of substrate entering the reactor (L h^{-1})
F_j	transfer function of the jth layer of the net ()
h	number of learning epoch
H_i	specific ionic constant of ion i (i = Na, Ca, Mg, Cl, CO$_3$, etc.)
I	input data from the training set
I_i	ionic strength of ion i (i = Na, Ca, Mg, Cl, CO$_3$, etc.)
k	discrete time

$(k_l a)$ product of mass-transfer coefficient for oxygen and gas-phase specific area (h^{-1})

$(k_l a)_0$ product of mass-transfer coefficient at 20 °C for O_2 and gas-phase specific area (h^{-1})

K_{O_2} constant of oxygen consumption (g L^{-1})

K_P constant of growth inhibition by ethanol (g L^{-1})

K_{P1} constant of fermentation inhibition by ethanol (g L^{-1})

K_S constant in the substrate term for growth (g L^{-1})

K_{S1} constant in the substrate term for ethanol production (g L^{-1})

K_T heat transfer coefficient (J h^{-1} m^{-2} K^{-1})

l_r learning rate

m momentum parameter used in the learning algorithm (0.95)

m_i quantity of inorganic salt i (i = NaCl, $CaCO_3$, $MgCl_2$) (g)

M_i molecular/atomic mass of salt/ion i (g mol^{-1})

N number of layers in the neural network (input layer is not counted)

P prediction horizon

Q number of sets of training input-output data

R universal gas constant (8.31 J mol^{-1} K^{-1})

r_{O_2} rate of oxygen consumption (mg L^{-1} h^{-1})

R_{SP} ratio of ethanol produced per glucose consumed for fermentation

R_{SX} ratio of cell produced per glucose consumed for growth

S_j number of neurons in the jth layer

t time (h)

T_{ag} temperature of cooling agent in the jacket (°C)

T_{in} temperature of the substrate flow entering to the reactor (°C)

$T_{in,ag}$ temperature of cooling agent entering to the jacket (°C)

T_r temperature in the reactor (°C)

T_{sp} setpoint temperature (°C)

U vector of the manipulated variables in linear models

V volume of the mass of reaction (L)

V_j volume of the jacket (L)

$n(i,j)^{(h)}$ weighting factor from the ith input variable to the jth output variable in the hth learning epoch

X state vector in linear models

Y output vector in linear models

Y_{O_2} yield factor for biomass on oxygen (mg mg^{-1}), defined as the amount of oxygen consumed per unit biomass produced

z ionic charge of ion i

Greek Symbols

$\delta w(i,j)^{(h)}$ variation of the weighting factor in the hth learning epoch

ΔH_r reaction heat of fermentation (kJ mol^{-1} O_2 consumed)

Φ, Γ discrete state space matrices used in linear models

μ_{O_2} maximum specific oxygen consumption rate (h^{-1})

μ_P maximum specific fermentation rate (h^{-1})

μ_X maximum specific growth rate (h^{-1})

ϱ_{ag} density of cooling agent $(g\ L^{-1})$

ϱ_r density of the mass of reaction $(g\ L^{-1})$

3.3
Dynamic Modeling and Control of a High-Purity Distillation Column

3.3.1
Introduction

Distillation is a common, but energy-intensive, method of performing separations in the petroleum, chemical, food, pulp and paper, and pharmaceutical industries. In the chemical and petroleum industries alone, distillation is used to effect 95 % of all separations. The thermal energy requirements of distillation are enormous. Distillation processes are thermodynamically less than 10 % efficient and account for approximately 8 % of the total energy use of the U. S. industrial sector. The U. S. Department of Energy's Office of Industrial Technologies (OIT) has targeted improved distillation as one of its major goals for saving energy in the industrial sector. One widely discussed way to save energy in distillation is to use an appropriate advanced control system, which is able to incorporate optimization algorithms in order to minimize to operation cost of the column.

It is well known that high-purity distillation columns are difficult to control due to their strong nonlinear behavior and the high interacting nature of the composition control loops. Usually, distillation columns are operated within a wide range of feed compositions and flow rates that make a control design even more difficult. However, a tight control of both product compositions is necessary to guarantee the smallest possible energy consumption, as well as high and uniform product qualities.

Model Predictive Control (MPC) is an important technique used in the process control industries [67,68]. MPC has been developed considerably during the past few years, due mainly to its multivariable nature that permits control problems to be addressed globally. MPC also permits plant constraints to be taken into account. MPC algorithms rely on a process model and on-line solution of a usually constrained optimization problem that minimizes a certain objective over a future horizon. For many applications, one of the major limitations of MPC techniques in their original formulation derives from the use of a linear model of the process inside the algorithm. Although linear model predictive control (LMPC) techniques have been implemented successfully in a wide variety of nonlinear systems, the use of these control strategies for certain nonlinear processes and operating conditions, is subject to performance limitations. A direct extension of the LMPC methods results when a nonlinear dynamic process model is used, rather than the linear convolution model. During the past decade, much effort has been invested to extend the model-based predictive approaches from linear systems to nonlinear ones. In this regard, three major issues must be considered in the extension of LMPC to nonlinear systems, namely, the stability of the closed-loop system, the computational burden, and modeling issues. The impacts of these challenges on nonlinear model predictive control (NMPC) implementation have been discussed widely in the literature [69–74], and in chapters 1 and 2 of the book. Moreover, several applications of NMPC to different distillation systems have been reported [75–78].

Since a constrained nonconvex nonlinear optimization problem must be solved on-line, the major practical challenge associated with NMPC is the computational complexity that increases significantly with the complexity of the models used in the controller. Additionally, significant progress has been made in the field of dynamic process optimization. Rapid on-line optimization algorithms have been developed that exploit the specific structure of optimization problems arising in NMPC, and real-time applications have been proven to be feasible for small-scale processes. However, the global solution of the optimization cannot be guaranteed. With respect to these obstacles in real-time implementation, it was pointed out that the feasibility implies stability in the case of QIH-NMPC and the global solution of the optimization problem can be relaxed [79].

Although the complexity of the first principle model used in NMPC represents the main obstacle in real-time implementation of these control approaches, until now there is no realistic feasibility study that discusses the computational burden vis-à-vis the model complexity and modeling difficulties in an application realistic scenario. In this chapter, it is shown that modern NMPC schemes in combination with specialized dynamic optimization strategies can be feasible even for large-scale process models. In particular, the control of a high-purity distillation column for the separation of methanol and n-propanol is considered, and the real-world implementation complexity and control performance are compared for models of different order. Additionally, techniques based on genetic algorithms (GA) – which makes solution of the optimization more robust – are presented. The implementation issues of artificial neural network (ANN)-based controllers for the simulated system are also discussed. The advantages and disadvantages of ANN-based controllers to the first principle model-based NMPC are highlighted, and a new adaptive ANN-based control approach is introduced.

3.3.2
Dynamic Modeling of the Binary Distillation Column

The column considered in our simulations has $N = 40$ trays, and is used for the separation of methanol and n-propanol (Figure 3.51).

The feed flow (F) enters into the column at tray 21, and is considered to be the main source of disturbance through the feed flow composition (x_F). The feed stream separates the column into the rectifying and stripping sections. Products are removed continuously at the bottom and top of the column with feed B, concentration x_B and feed D and concentration x_D, respectively. The vapor phase is condensed totally in the condenser and is partly fed back into the column on the last tray with the reflux flow (L_{N+1}). The bottom stream is introduced into the reboiler where part of it is vaporized and reintroduced on the first tray with flow rate V_D, and

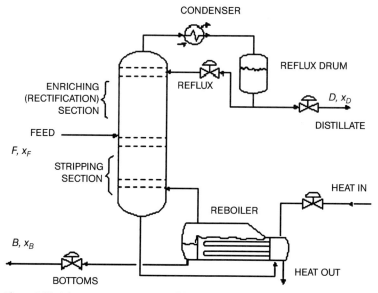

Figure 3.51 Schematic representation of the distillation column.

the remainder is removed as bottom product (B). In both, the condenser and reboiler, a perfect level control is considered. The column is considered in LV configuration – that is, L_{N+1} and V_0 are considered as the control inputs.

In order to estimate the feasibility of NMPC for real-time process control with regard to model complexity, different models have been developed and used in the controller for comparison. In each model, the following general assumptions were considered, in addition to which in each model some particular considerations are additionally used:

- A1: two-phase system in thermal and mechanical equilibrium;
- A2: perfect mixing in vapor and liquid phases;
- A3: no heat of mixing;
- A4: no heat loss to the surroundings;
- A5: temperature dynamics of the column structure is neglected;
- A6: thermodynamic equilibrium of vapor and liquid phases;
- A7: constant liquid holdups ($dn_j/dt = 0$); and
- A8: vapor holdup is neglected ($n_j^V = 0 \Rightarrow n_j = n_j^V + n_j^L \approx n_j^L$);

The modeling of distillation columns has been widely discussed in the literature, and some excellent reviews have been published [80–83]. In this section, the five models of different complexity are presented briefly. Models A, B, and C are obtained from basic conservation laws and constitutive relations written for the trays represented schematically in Figure 3.52.

Models D and E are derived from model C, using a specific model reduction technique.

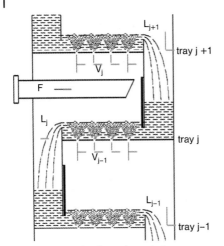

Figure 3.52 Molar flows through the trays.

3.3.2.1 Model A: 164th Order DAE Model

Model A is obtained by "rigorous" modeling of the distillation system. Considering assumptions A1 to A8, a complex DAE model, involving 42 differential $(x_j, j = \overline{0, N+1}$ with $x_0=x_B$ and $x_{N+1}=x_D)$ and 122 algebraic states $(L_j, V_j, T_j, j = \overline{1, N}, T_0 = T_B$ and $T_{N+1} = T_D)$ was derived. The equations describing model A are as follows:

- Reboiler: $(j = 0)$

$$\frac{dx_B}{dt} = \frac{1}{n_B}(-B \cdot x_B - V_B \cdot y_B + L_1 \cdot x_1) \tag{3.203}$$

$$0 = -V_B + L_1 - B \Rightarrow B = L_1 - V_B \tag{3.204}$$

- Condenser: $(j = N+1)$

$$\frac{dx_D}{dt} = \frac{1}{n_D} \cdot (V_N \cdot y_N - D \cdot x_D - L_{N+1} \cdot x_D) \tag{3.205}$$

$$0 = V_N - D - L_{N+1} \Rightarrow D = V_N - L_{N+1} \tag{3.206}$$

- Tray j: $(j = \overline{1, N}$ and $j \neq M)$

$$\frac{dx_j}{dt} = \frac{1}{n_j}(V_{j-1} \cdot y_{j-1} - V_j \cdot y_j + L_{j+1} \cdot x_{j+1} - L_j \cdot x_j) \tag{3.207}$$

$$0 = V_{j-1} - V_j + L_{j+1} - L_j \tag{3.208}$$

$$0 = V_{j-1} \cdot h_{j-1}^V - V_j \cdot h_j^V + L_{j+1} \cdot h_{j+1}^L - L_j \cdot h_j^L - n_j \cdot \frac{dx_j}{dt} \cdot \left(h_{1,j}^L - h_{2,j}^L \right) \tag{3.209}$$

- Feed tray ($j = M$)

$$\frac{dx_M}{dt} = \frac{1}{n_M}(V_{M-1} \cdot y_{M-1} - V_M \cdot y_M + L_{M+1} \cdot x_{M+1} - L_M \cdot x_M + F \cdot x_F) \tag{3.210}$$

$$0 = V_{M-1} - V_M + L_{M+1} - L_M + F \tag{3.211}$$

$$0 = V_{M-1} \cdot h_{M-1}^V - V_M \cdot h_M^V + L_{M+1} \cdot h_{M+1}^L - L_M \cdot h_M^L -$$

$$n_M \cdot \frac{dx_M}{dt} \cdot \left(h_{1,M}^L - h_{2,M}^L\right) + F \cdot h_F^L + (1 - q_F) \cdot F \cdot h_F^V \tag{3.212}$$

Remark: *The energy balance used in the equations was derived as follows:*
From the general equation:

$$\frac{dU_j}{dt} = V_{j-1} \cdot h_{j-1}^V - V_j \cdot h_j^V + L_{j+1} \cdot h_{j+1}^L - L_j \cdot h_j^L \text{ with}$$

$$U_j = n_j^L \cdot h_j^L + \underbrace{n_j^V \cdot h_j^V}_{=0,(A8)} = n_j^L \cdot h_j^L = n_j h_j^L \text{ one can write:}$$

$$\frac{dn_j h_j^L}{dt} = n_j \cdot \frac{dh_j}{dt} + h_j \cdot \underbrace{\frac{dn_j}{dt}}_{=0,(A7)} = n_j \cdot \frac{dh_j}{dt} = n_j \cdot \frac{dh_j^L}{dx_j} \cdot \frac{dx_j}{dt} =$$

$$= n_j \cdot \frac{d\left(x_j \cdot h_{1,j}^L + (1 - x_j) \cdot h_{2,j}^L\right)}{dx_j} \cdot \frac{dx_j}{dt} = n_j \cdot \left(h_{1,j}^L - h_{2,j}^L\right) \cdot \frac{dx_j}{dt}$$

The thermodynamic equilibrium is modeled by Raoult's law:

$$y_j = \frac{p_{1,j}}{p_j} \cdot x_j \tag{3.213}$$

The dependence of the vapor pressure on the temperature is described by Antoine's relationship:

$$p_{k,j} = \exp\left(A_k - \frac{B_k}{C_k + T_j}\right) \text{ for } k = 1, 2 \tag{3.214}$$

The total pressure at each tray is:

$$p_j = p_{1,j} \cdot x_j + p_{2,j} \cdot (1 - x_j) \tag{3.215}$$

Constant pressure drop between trays is considered:

$$p_j = p_{j-1} - \Delta p \tag{3.216}$$

The enthalpies of the liquid and vapor phases are computed as functions of temperature, pressure and composition of each tray, using the following equations:

$$h_j^L = x_j \cdot h_{1,j}^L + (1 - x_j) \cdot h_{2,j}^L \tag{3.217}$$

$$h_j^V = y_j \cdot h_{1,j}^V + (1 - y_j) \cdot h_{2,j}^V \tag{3.218}$$

$$h_{k,j}^L = a_k \cdot T_j + b_k \cdot T_j^2 + c_k \cdot T_j^3 \text{ for } k = 1, 2 \tag{3.219}$$

$$h_{k,j}^V = h_{k,j}^L + \Delta h_{k,j}^v(p_j, T_j) \text{ for } k = 1, 2 \tag{3.220}$$

$$\Delta h_{k,j}^v(p_j, T_j) = R \cdot Tc_k \cdot \sqrt{1 - \frac{Pr_{k,j}}{Tr_{k,j}^3}} \cdot (6.09648 - 1.28862 \cdot Tr_{k,j} + \newline +1.016 \cdot Tr_{k,j}^7 + \beta_k \cdot (15.6875 - 13.4721 \cdot Tr_{k,j} + 2.615 \cdot Tr_{k,j}^7)) \quad \text{for } k=1,2 \tag{3.221}$$

$$Tr_{k,j} = \frac{T_j}{Tc_k} \text{ and } Pr_{k,j} = \frac{p_j}{Pc_k} \tag{3.222}$$

These equations describe a complex 164th order DAE model with 42 differential and 122 algebraic states. The control variables in this case are:

$$L_c = L_{N+1} \text{ and } V_c = V_B$$

3.3.2.2 Model B: 84th Order DAE Model

Model B denotes a model that, besides assumptions A1 to A8, considers constant molar flows (L_j = const. and V_j = const., for all $j = \overline{0, N}$). Thus, the energy balance is not needed. From the total mass balance written for the feed tray one has:

$$0 = V_{M-1} - V_M + L_{M+1} - L_M + F \quad \Rightarrow \quad L_M = L_{M+1} + F \tag{3.223}$$

Thus, the differential equations of the model can be obtained from Eqs. (3.203) to (3.207) and (3.213) by using the following notations:

$$V_j = V, \quad \text{for } j = \overline{0, N} \newline L_j = \begin{cases} L & \text{for } j = \overline{M+1, N+1} \\ L+F & \text{for } j = \overline{1, M} \end{cases} \tag{3.224}$$

In order to obtain the composition of the vapor phase (y_j) and the temperatures (T_j) on each stage, the same system of algebraic equations, (3.213) to (3.216), is written for every tray, condenser and reboiler. Thus, an 84th order DAE model with 42 differential ($x_j, j = \overline{0, N+1}$) and 42 algebraic states ($T_j, j = \overline{1, N}, T_0 = T_B$ and $T_{N+1} = T_D$) is derived with the control inputs: $L_c = L$ and $V_c = V$.

3.3.2.3 Model C: 42nd Order ODE Model

Model C was derived from model B, by introducing additional simplifying assumptions. Besides the constant molar overflows and neglecting energy balances, constant pressure and constant relative volatility are considered. Hence, model C is described by the same differential equations as model B. The composition of the vapor phase is obtained under the assumption of constant relative volatility ($\alpha_r = $ - *const.*), from the following equation:

$$y_j = \frac{\alpha_r \cdot x_j}{1 + (\alpha_r - 1) \cdot x_j} \quad \text{for } j = \overline{0, N+1} \tag{3.225}$$

These equations lead to a 42nd order ODE model with the same 42 differential states $(x_j, j = \overline{0, N+1})$ and 2 control inputs ($L_c = L$ and $V_c = V$) as in model B. Although this is the simplest model from a modeling point of view, it can still reproduce the essential dynamic behavior of the system.

3.3.2.4 Model D: 5th Order ODE Model

Model D uses the same assumptions as model C, and was derived from the latter by using the so-called "nonlinear wave propagation" theory [84,85]. This theory is based on the property of the concentration profiles in a distillation column having a stable form (different for the stripping and rectifying sections), which moves along the column when disturbances appear. The steady-state concentration profile can be described by the following wave function:

$$x(z) = \mu_- + \frac{\mu_+ - \mu_-}{1 + e^{-\varrho(z-s-\xi)}} \tag{3.226}$$

The shape parameters ϱ and ξ are determined for each section, with a least-squares fitting method, such that the concentration profile matches the values obtained with the function. The parameters μ_- and μ_+ represent the asymptotic limits of $x(z)$, for $z \rightarrow \pm\infty$, and can be determined from the system of equations written for the boundary system (reboiler/feed tray for the stripping section and feed tray/condenser for the rectifying section):

$$\begin{cases} x_{z_0} - \mu_- + \dfrac{\mu_+ - \mu_-}{1 + e_{z_0}} & \text{with, } e_{z_0} = e^{-\varrho(z_0 - s - \xi)} \\ x_{z_N} = \mu_- + \dfrac{\mu_+ - \mu_-}{1 + e_{z_N}} & \text{with, } e_{z_N} = e^{-\varrho(z_N - s - \xi)} \end{cases} \tag{3.227}$$

with the solution:

$$\mu_- = \frac{x_{z_N} \cdot (1 + e_{z_N}) - x_{z_0} \cdot (1 + e_{z_0})}{e_{z_N} - e_{z_0}} \tag{3.228}$$

$$\mu_+ = \frac{x_{z_0} \cdot e_{z_N} \cdot (1 + e_{z_0}) - x_{z_N} \cdot e_{z_0} \cdot (1 + e_{z_N})}{e_{z_N} - e_{z_0}} \tag{3.229}$$

Variables x_{z_0} and x_{z_N} represent the boundary concentration values for each section of the column as follows:
- Stripping section:

$$x_{z_0} = x_B \quad \text{for } z_0 = 0 \text{ (reboiler)}$$

$$x_{z_N} = x_M \quad \text{for } z_N = M \text{ (feed tray)}$$

- Rectifying section:

$$x_{z_O} = x_M \quad \text{for } z_O = M \text{ (feed tray)}$$

$$x_{z_N} = x_D \quad \text{for } z_N = N + 1 \text{ (condenser)}$$

By using Eqs. (3.226), (3.228), and (3.229), the concentrations on every tray (for each section) can be obtained easily as a function of z, s, x_{z_O} and x_{z_N} for $z = z_O + 1, z_N - 1$.

The differential equation, which gives the position of the inflection point (s) of the concentration profile, is obtained by a summation of the material balances over all trays in one section:

$$\sum_{z=z_O+1}^{z_N-1} n_z \cdot \frac{dx(z)}{dt} = L \cdot (x_{z_N} - x_{z_O+1}) - V \cdot (y_{z_N-1} - y_{z_O}) \qquad (3.230)$$

with

$$L = \begin{cases} L & \text{for rectifying section} \\ L + F & \text{for stripping section} \end{cases} \qquad (3.231)$$

By substituting in Eq. (3.230) the following equation:

$$\frac{dx(z)}{dt} = \frac{dx(z)}{ds} \cdot \frac{ds}{dt} + \frac{dx(z)}{dx_{z_O}} \cdot \frac{dx_{z_O}}{dt} + \frac{dx(z)}{dx_{z_N}} \cdot \frac{dx_{z_N}}{dt} \qquad (3.232)$$

one can obtain a complicated expression which provides the differential equation for s. The derivatives of $x(z)$ with respect to s, x_{z_O} and x_{z_N} can be obtained from Eqs. (3.226), (3.228), and (3.229).

The equations above finally lead to a 5th order ODE model with the boundary concentrations (x_B, x_M and x_D) and the turning points of the concentration profiles for the two sections (s_s for the stripping section and s_r for the rectifying section) as differential states. Concentrations x_B, x_M and x_D are obtained from the same equations as in the case of model C.

3.3.2.5 Model E: 5th Order DAE Model

One drawback of model D is the stiffness of the ODEs, caused by the parameters with very fast and very slow dynamics. It can be easily shown that the three concentrations (x_B, x_M and x_D) have a much faster dynamic behavior than the wave positions (s_s and s_r). Thus, another model can be derived from model D using the quasi steady-state assumption for the three concentrations with fast dynamics [86]. The resulting model E is a 5th order DAE model with three algebraic (x_B, x_M and x_D) and two differential (s_s and s_r) variables described by the same equations as model D, but with the right-hand side of Eqs. (3.226), (3.228), and (3.229) set equal to zero.

3.3.2.6 Comparison of the Models

For a better overview of the complexity of the five models, the numbers of differential and algebraic variables used in each model are summarized in Table 3.6. The most important disturbance that can appear in the system is variation of the feed composition. Therefore, the dynamic and steady-state behavior of the distillation column, as described by the five different models, are studied for step changes in the feed composition. Figure 3.53 shows the steady-state concentration profiles as well as the variation of the liquid and vapor flows in the case of model A, compared to the constant molar flows used in all other models, before and after a step +20% in the feed composition in x_F. One can observe that models C, D, and E have very similar steady-state behaviors. This is explained by the identical modeling assumptions used for these models. Models A, B, and C describe the system quite differently due to the different levels of complexity and assumptions used. An interest-

Table 3.6 Overview of model complexity.

Model	Order and type	No. of differential states	No. of algebraic states
A	164th DAE	42	122
B	64th DAE	42	42
C	42nd ODE	42	0
D	5th ODE	5	0
E	5th DAE	2	3

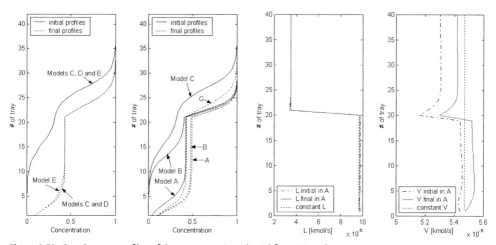

Figure 3.53 Steady-state profiles of the concentrations, liquid flows [L] and vapor flows [V] for +20% step in x_F.

ing conclusion that can be drawn from Figure 3.53 is that, in the case of disturbance in the feed concentration, constant liquid flow seems to be a valid assumption, but not constant vapor flow. Figure 3.54 shows the results for a –20% step change in x_F. These figures confirm the conclusions drawn above.

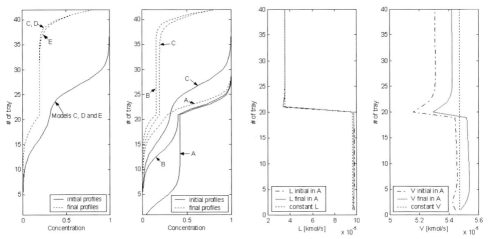

Figure 3.54 Steady-state profiles of the concentrations, liquid flows [L] and vapor flows [V] for –20% step in x_F.

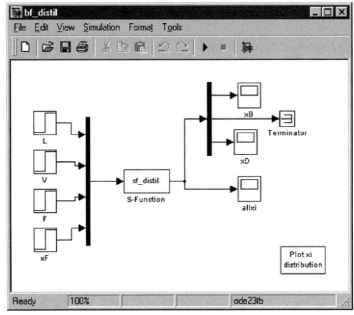

Figure 3.55 SIMULINK block diagram used for simulation.

The models were implemented as MATLAB S-functions, and simulations were carried out using the SIMULINK block diagram shown in Figure 3.55. The dynamic response of the system (obtained with model C) for different step changes in the feed flow rate and composition, respectively, is shown in Figures 3.56–3.60.

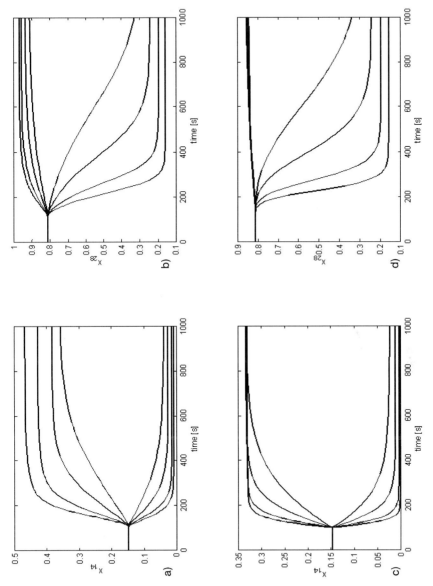

Figure 3.56 Variation of liquid concentration on trays 14 and 28: (a,b) in the case of step changes in the feed flow concentration by ±30%, ±20%, ±10% and ±5%; (c,d) in the case of step changes in the feed flow rate by by ±30%, ±20%, ±10% and ±5%.

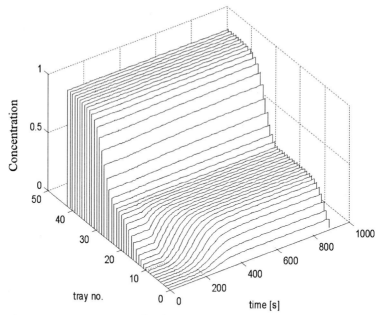

Figure 3.57 Concentration profiles for +20% step in the feed
flow rate at $t = 100$ s.

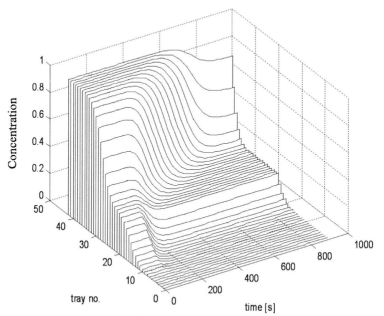

Figure 3.58 Concentration profiles for −20% step in the feed
flow rate at $t = 100$ s.

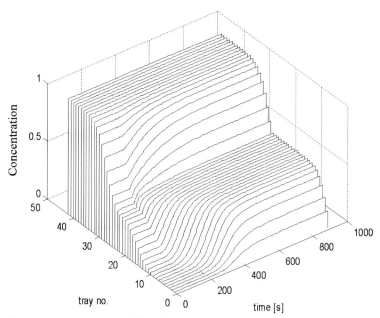

Figure 3.59 Concentration profiles for +30% step in the feed
flow composition at $t = 100$ s.

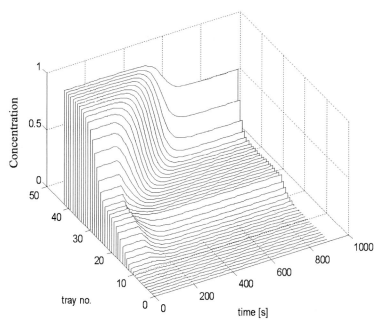

Figure 3.60 Concentration profiles for –30% step in the feed
flow composition at $t = 100$ s.

3.3.3
A Computational Efficient NMPC Approach for Real-Time Control of the Distillation Column

3.3.3.1 NMPC with Guaranteed Stability of the Distillation Column

The objective of NMPC is to calculate a set of future control moves (control horizon, T_c) by minimization of a cost function, such as the squared control error on a moving finite horizon (prediction horizon, T_p). The optimization problem is solved on-line based on predictions obtained from a nonlinear model. Here, we reformulate in a more general form the NMPC problem based on the first principle models described in Section 3.3.2. The nonlinear model used for prediction will be represented by a set of differential and algebraic equations (DAE):

$$\dot{\mathbf{x}}(t) = \mathbf{f}(\mathbf{x}(t), \mathbf{z}(t), \mathbf{u}(t)), \qquad \mathbf{x}(0) = \mathbf{x_0} \tag{3.233}$$
$$0 = \mathbf{g}(\mathbf{x}(t), \mathbf{z}(t), \mathbf{u}(t))$$

with differential variables $\mathbf{x}(t) \in \mathbb{R}^n$, algebraic variables $\mathbf{z}(t) \in \mathbb{R}^p$ and inputs $\mathbf{u}(t) \in \mathbb{R}^m$. The setpoints are the steady-state values $(\mathbf{x}_s, \mathbf{z}_s, \mathbf{u}_s)$ satisfying $0 = \mathbf{f}(\mathbf{x}_s, \mathbf{z}_s, \mathbf{u}_s)$ and $0 = \mathbf{g}(\mathbf{x}_s, \mathbf{z}_s, \mathbf{u}_s)$. Because of safety considerations and physical limitations, it is additionally often required that constraints of the form $\mathbf{h}(\mathbf{x}_s, \mathbf{z}_s, \mathbf{u}_s) \geq 0$ need to be satisfied. Additionally, another important requirement that the nonlinear model predictive controllers must meet is that they should assure a stable closed-loop system. The implementation issues of two FPNMPC schemes, which guarantee nominal stability, are discussed in this section.

The most widely suggested stability constraint is the terminal equality constraint, which forces the states to be zero (equal to their steady-state values) at the end of the finite horizon. For a certain steady-state this condition is written:

$$\bar{\mathbf{x}}(t + T_P) - \mathbf{x}_s = 0 \tag{3.234}$$

Using the terminal equality constraint to guarantee stability is an intuitive approach; however, it increases significantly the on-line computation necessary to solve the open-loop optimization problem and often causes feasibility problems.

The second NMPC approach to guarantee stability discussed here is the quasi-infinite horizon nonlinear MPC, in which the prediction horizon is extended approximately to infinity by introducing a terminal penalty term in the objective function. The difference between this approach and the one described above is that, rather than using a terminal equality constraints, a terminal region is used and the objective function has an additional term, the terminal penalty term. The key problem in this approach is the choice of the form, and the off-line computation of the terminal region and penalty term, which in general – because of the nonlinearity of the system – is a difficult task. In the case of using a local linear feedback law and a quadratic objective function, the terminal penalty term can be chosen to be quadratic. In this case, the terminal penalty matrix Q_p is the solution of a Lyapunov equation, and the on-line control problem can be expressed as below for a DAE system:

$$\min_{\mathbf{u}(\cdot)} \{J_s(\overline{\mathbf{x}}(\cdot), \overline{\mathbf{u}}(\cdot))\} \tag{3.235}$$

with

$$J_s(\overline{\mathbf{x}}(\cdot), \overline{\mathbf{u}}(\cdot)) \overset{def}{=} \underbrace{\int_t^{t+T_p} \left(\|\overline{\mathbf{x}}(\tau) - \mathbf{x}_s\|_Q^2 + \|\overline{\mathbf{u}}(\tau) - \mathbf{u}_s\|_R^2 \right) d\tau}_{\text{Lagrange term}} + \underbrace{E_s(\overline{\mathbf{x}}(t + T_p))}_{\text{Mayer term}} \tag{3.236}$$

subject to:

$$\dot{\mathbf{x}} = \mathbf{f}(\overline{\mathbf{x}}(\tau), \mathbf{z}(\tau), \overline{\mathbf{u}}(\tau)), \quad \overline{\mathbf{x}}(t) = \mathbf{x}(t) \tag{3.237}$$

$$0 = \mathbf{g}(\overline{\mathbf{x}}(\tau), \mathbf{z}(\tau), \overline{\mathbf{u}}(\tau)), \quad \tau \in [t, t + T_p] \tag{3.238}$$

$$\mathbf{h}(\overline{\mathbf{x}}(\tau), \mathbf{z}(\tau), \overline{\mathbf{u}}(\tau)) \geq \mathbf{0}, \quad \tau \in [t, t + T_p] \tag{3.239}$$

$$r(\overline{\mathbf{x}}(t + T_p)) \geq 0 \tag{3.240}$$

where $\|x\|_Q^2 = x^T \cdot Q \cdot x$, with Q a positive-definite matrix, and the bar indicates that the corresponding variables are used in the controller. In the cost function, matrices Q and R weight the state and input deviation from their operating points. The terminal penalty term is given by the following expression:

$$E_s(\overline{\mathbf{x}}) = \|\mathbf{x} - \mathbf{x}_s\|_{Q_p}^2 \tag{3.241}$$

The terminal penalty matrix Q_p weights the terminal state deviation from the steady-state values, and the terminal constraint restricts the predicted state at the end of the prediction horizon to lie in an ellipsoid (Ω) around the setpoint:

$$r(\overline{\mathbf{x}}) = \alpha - E_s(\overline{\mathbf{x}}) \tag{3.242}$$

Note that if the steady state of the system changes during the operation (for example, due to step changes in the input parameters), the terminal region and the terminal penalty matrix Q_p, must be adjusted to the new steady-state values.

The procedure to obtain the terminal penalty matrix and the terminal region for systems described by ODE (see Section 3.4) or DAE models as well as the setpoint tracking technique are presented in the literature [70,87].

In order to obtain some algorithmic advantages in solving the optimization from the control problem, model parameters p are introduced as variable. The method presented next was first proposed and described by Diehl et al. [88]. According to this approach, the NMPC problem [Eqs. (3.235) to (3.240)], is rewritten in a more general form:

$$\min_{\mathbf{u}(\cdot), \mathbf{x}(\cdot), \mathbf{p}} \{J_s(\overline{\mathbf{x}}(\cdot), \overline{\mathbf{z}}(\cdot), \overline{\mathbf{u}}(\cdot), \overline{\mathbf{p}})\} \tag{3.243}$$

with

$$J_s\left(\bar{\mathbf{x}}(\cdot),\bar{\mathbf{z}}(\cdot),\bar{\mathbf{u}}(\cdot),\bar{\mathbf{p}}\right) \overset{def}{=} \int\limits_t^{t+T_p} L\left(\bar{\mathbf{x}}(\tau),\bar{\mathbf{z}}(\tau),\bar{\mathbf{u}}(\tau),\bar{\mathbf{p}}\right)d\tau + E_s\left(\bar{\mathbf{x}}(t+T_p),\bar{\mathbf{p}}\right) \tag{3.244}$$

subject to:

$$\dot{\mathbf{x}} = \mathbf{f}\left(\bar{\mathbf{x}}(\tau),\bar{\mathbf{z}}(\tau),\bar{\mathbf{u}}(\tau),\bar{\mathbf{p}}\right) \tag{3.245}$$

$$\bar{\mathbf{x}}(t) = \mathbf{x}(t) \tag{3.246}$$

$$\bar{\mathbf{p}} = \mathbf{p}(t) \tag{3.247}$$

$$0 = \mathbf{g}\left(\bar{\mathbf{x}}(\tau),\bar{\mathbf{z}}(\tau),\bar{\mathbf{u}}(\tau),\bar{\mathbf{p}}\right), \quad \tau \in \left[t, t+T_p\right] \tag{3.248}$$

$$\mathbf{h}\left(\bar{\mathbf{x}}(\tau),\bar{\mathbf{z}}(\tau),\bar{\mathbf{u}}(\tau),\bar{\mathbf{p}}\right) \geq 0, \quad \tau \in \left[t, t+T_p\right] \tag{3.249}$$

$$r\left(\bar{\mathbf{x}}(t+T_p),\bar{\mathbf{p}}\right) \geq 0 \tag{3.250}$$

3.3.3.2 Direct Multiple Shooting Approach for Efficient Optimization in Real-Time NMPC

3.3.3.2.1 Parameterization of the Optimization Problem

Using direct multiple shooting, the open-loop optimization in problem in the NMPC at each time t for the current system state $\mathbf{x}(t)$ can be solved in the following way:

- First, the predicted control trajectory $\mathbf{u}(\cdot)$ is discretized. In this case it is assumed that the controls are piecewise constant on each of the $N = Tp/\delta$ predicted sampling intervals:

$$\bar{\mathbf{u}}(\tau) = \bar{\mathbf{u}}_i \ \text{ for } \ \tau \in [\tau_i, \tau_{i+1}), \quad \tau_i = t + i \cdot \delta \tag{3.251}$$

- Second, the DAE solution is decoupled on these intervals by considering the initial values $\bar{s}^x_i$ and $\bar{s}^z_i$ of differential and algebraic states at the times τ_i as additional optimization variables. The solution of such a decoupled initial value problem is denoted by $\bar{\mathbf{x}}_i(\cdot)$, $\bar{\mathbf{z}}_i(\cdot)$; it obeys the following relaxed DAE formulation on the interval $[\tau_i, \tau_{i+1})$:

$$\dot{\mathbf{x}}_i(\tau) = \mathbf{f}\left(\bar{\mathbf{x}}_i(\tau),\bar{\mathbf{z}}_i(\tau),\bar{\mathbf{u}}_i,\bar{\mathbf{p}}\right) \tag{3.252}$$

$$0 = \mathbf{g}\left(\bar{\mathbf{x}}_i(\tau),\bar{\mathbf{z}}_i(\tau),\bar{\mathbf{u}}_i,\bar{\mathbf{p}}\right) - \alpha_i(\tau) \cdot \mathbf{g}\left(\bar{s}^x_i,\bar{s}^z_i,\bar{\mathbf{u}}_i,\bar{\mathbf{p}}\right) \tag{3.253}$$

$$\bar{\mathbf{x}}_i(\tau_i) = \bar{s}^x_i, \quad \bar{\mathbf{z}}_i(\tau) = \bar{s}^z_i \tag{3.254}$$

The subtrahend in Eq. (3.253) is deliberately introduced to allow an efficient DAE solution for initial values and controls $\bar{s}_i^x$, $\bar{s}_i^z$, $\bar{u}_i$ that violate temporarily the consistency conditions $0 = g(\bar{s}_i^x, \bar{s}_i^z, \bar{u}_i, \bar{p})$, during the solution iterations [89–91]. Therefore, we require for the scalar damping factor α that $\alpha_i(t_i) = 1$. Note that the trajectories $x_i(\tau)$ and $z_i(\tau)$ are functions of the initial values, controls and parameters: $\bar{s}_i^x$, $\bar{s}_i^z$, $\bar{u}_i$, $\bar{p}$. The objective contribution of the Lagrange term on $[\tau_i, \tau_{i+1})$ is – like the DAE solutions $\bar{x}_i(\tau_i)$, $\bar{z}_i(\tau)$ – completely determined by $\bar{s}_i^x$, $\bar{s}_i^z$, $\bar{u}_i$, $\bar{p}$:

$$J_{s,i}\left(\bar{s}_i^x, \bar{s}_i^z, \bar{u}_i, \bar{p}\right) = \int_{\tau_i}^{\tau_{i+1}} \left(\|\bar{x}(\tau) - x_s\|_Q^2 + \|\bar{u}(\tau) - u_s\|_R^2\right) \cdot d\tau \tag{3.255}$$

3.3.3.2.2 The Structured Nonlinear Programming (NLP) Problem

The parametrization of problem of Eqs. (3.243) to (3.250) using multiple shooting and a piecewise constant control representation leads to the following large, but specially structured, NLP problem:

$$\text{Solve: } \min_{u_i, s_i, p} \sum_{i=0}^{N-1} J_{s,i}\left(\bar{s}_i^x, \bar{s}_i^z, \bar{u}_i, \bar{p}\right) + E_s\left(\bar{s}_N^x, \bar{p}\right) \tag{3.256}$$

$$\text{Subject to: } \bar{s}_0^x = x(t), \tag{3.257}$$

$$\bar{p} = p(t) \tag{3.258}$$

$$\bar{s}_{i+1}^x = \bar{x}_i(\tau_{i+1}), \quad i = 0, 1, \ldots N-1, \tag{3.259}$$

$$0 = g\left(\bar{s}_i^x, \bar{s}_i^z, \bar{u}_i, \bar{p}\right), \quad i = 0, 1, \ldots N, \tag{3.260}$$

$$h\left(\bar{s}_i^x, \bar{s}_i^z, \bar{u}_i, \bar{p}\right) \geq 0, \quad i = 0, 1, \ldots N, \tag{3.261}$$

$$r_s\left(\bar{s}_N^x, \bar{p}\right) \geq 0. \tag{3.262}$$

This large structured NLP problem in the variables $(\bar{s}_0^x, \bar{s}_0^z, \bar{u}_0, \ldots)$ is solved by a specially tailored, *partially reduced* SQP algorithm [89,90,92,93].

3.3.3.2.3 SQP for Multiple Shooting

The sequential quadratic programming (SQP) method tailored to the multiple shooting structure of the NLP problem presented above [Eqs. (3.256) to (3.262)], can be expressed as follows:

• The NLP can be summarized as

$$\min_w F(w) \quad \text{subject to} \quad \begin{cases} G(w) = 0 \\ H(w) \geq 0 \end{cases} \tag{3.263}$$

where w contains all the multiple shooting state variables and also controls the model parameters. The discretized dynamic model is included in the equality constraints $G(w) = 0$.

Starting from an initial guess w^0, an SQP method for the solution of Eq. (3.263) iterates:

$$w^{k+1} = w^k + \alpha^k \cdot \Delta w^k, \quad k = 0, 1, \dots, \tag{3.264}$$

where $\alpha^k \in [0, 1]$ is a relaxation factor, and the search direction Δw^k is the solution of the quadratic programming (QP) subproblem:

$$\min_{\Delta w \in \Omega^k} \nabla F\left(w^k\right)^T \cdot \Delta w + \frac{1}{2} \cdot \Delta w^T \cdot A^k \cdot \Delta w \tag{3.265}$$

subject to:

$$\begin{aligned} G\left(w^k\right) + \nabla G\left(w^k\right)^T \cdot \Delta w &= 0 \\ H\left(w^k\right) + \nabla H\left(w^k\right)^T \cdot \Delta w &\geq 0 \end{aligned} \tag{3.266}$$

A^k denotes an approximation of the Hessian of the *Lagrangian function l*:

$$l(w, \lambda, \mu) = F(w) - \lambda^T \cdot G(w) - \mu^T \cdot H(w) \tag{3.267}$$

where λ and μ are the Lagrange multipliers.

Due to the choice of state and control parameterizations, the NLP problem and the resulting QP problems have a particular structure: the Lagrangian function l is partially separable – that is, it can be written in the form:

$$l(w, \lambda, \mu) = \sum_{i=0}^{N} l_i(w_i, \lambda_i, \mu_i) \tag{3.268}$$

where w_i are the components of the variables w corresponding to the interval $[t_i, t_i+1]$. This separation is possible if we simply interpret the parameters p as piecewise constant continuous controls. The Hessian of l, therefore has a *block diagonal structure* with blocks $\nabla^2_{w_i} l_i(w_i, \lambda, \mu)$. Similarly, the multiple shooting parameterization introduces a characteristic *block sparse structure* of the Jacobian matrices $\nabla G(w)^T$ and $\nabla H(w)^T$. A number of specific algorithmic developments were proposed by Diehl et al. [88], that exploit these specific features, leading to significant enhancement of the performance and numerical stability of the direct multiple shooting method. For example, the block diagonal structure of the Hessian, may be exploited in three different ways:

- In the first approach, a numerical calculation of the exact Hessian that corresponds to Newton's method is performed. This version is recommended if computation of the Hessian is cheap or if it can be computed and stored in advance.

- A second approach, that of partitioned high-rank updates (as introduced in Bock and Plitt [94]), speeds up local convergence with negligible computational effort for the Hessian approximation.
- A third approach for efficient Hessian approximation is the constrained Gauss–Newton method. This approach is especially recommended in the special case of a least squares type cost functions of the form $F(w) = 1/2 \cdot \|C(w)\|_2^2$. The matrix $\nabla_w C^T \cdot \nabla_w C$ is already available from the gradient computation and provides an excellent approximation of the Hessian, if the residual $C(w)$ of the cost function is sufficiently small.

An important point for using the above methods in the real-time context is their excellent local convergence behavior, which can be proven under certain assumptions. The aforementioned approaches were integrated in a state-of-the-art software package, MUSCOD-II, developed by the Interdisciplinary Center for Scientific Computing (IWR) at the University of Heidelberg, Germany. For a detailed description of globalization strategies available in the latest version of direct multiple shooting (MUSCOD-II), the reader is referred to [95]. However, some other techniques (based on GAs), which significantly increase the chance to obtain global optima, are presented later in this chapter.

In a real-time scenario the optimal control problems solved in the NMPC scheme are different at each progressing time t. The difference among the problems, besides time t, is in the initial values $x(t)$, which we expect to deviate from the values predicted by the model. We must also expect that some of the parameters $p(t)$, which are assumed to be constant in the model, are subject to disturbances. The time between consecutive optimization problems must be short enough to guarantee a sufficiently rapid reaction to disturbances. In real-time approaches it is usually assumed that a sequence of neighboring optimization problems is solved. Assuming that a solution of the optimization problem for values t, $x(t)$, $p(t)$ is available, including function values, gradients and a Hessian approximation, but that at time t the real values of the process are the deviated values $x'(t)$, $p'(t)$, the *conventional approach* to obtain an updated value for the feedback control $u(x'(t), p'(t), T_p)$ is: (i) to start the SQP procedure as described above from the deviated values $x'(t)$, $p'(t)$ and to use the *old* control values u_i for an integration over the complete interval $[t, t + T_p]$; and then (ii) to iterate until a given convergence criterion is satisfied.

Another approach, suggested by [93,96], has the following particular features. First, the solution iterations are started from the solution for the reference values $x(t)$, $p(t)$ instead of the deviated values. This approach has significant computational advantages. In this algorithm, the first iteration is available in a negligible fraction of the time of a whole SQP iteration. It can be shown that the error of this first correction in the NLP variables compared to the solution of the full nonlinear problem is only second order in the size of the disturbance.

Based on this observation, it is possible not to iterate the nonlinear solution procedure to convergence, but rather to use the following scheme:

- Apply the result of the first correction of the controls immediately to the real process during some process duration δ.

- During this period δ first compute the full QP solution and, based on this iterate, compute a new linear-quadratic model expansion for the next time step, and solve the QP as far as possible to prepare the immediate feedback response for the following step.

It must be mentioned here that this procedure is not equivalent to the multiple shooting SQP method as described above: in each step or iteration a different optimal control problem is treated, with different $\mathbf{x}(t)$, $\mathbf{p}(t)$ in the NMPC case, and additionally different T_p for shrinking time intervals as in batch processes. In contrast to SQP methods, the solution procedures must be modified; for example, the quadratic programs must be solved without knowledge of the unknown values $\mathbf{x}(t)$, $\mathbf{p}(t)$ generating a feedback law that includes state and control inequality constraints. The time δ required for model expansion and full QP solution depends on the complexity of the model and the optimization problem, as well as on the numerical solution algorithms involved and the available computer. If δ is not sufficiently small, parallelization may be a remedy.

3.3.3.3 Computational Complexity and Controller Performance

An application of the outlined NMPC schemes and direct multiple shooting method to control the high-purity distillation column described previously is provided in the next section. In particular, the computational and implementation complexity, as well as the resulting controller performance, are compared for different model sizes. First, the computational issues are considered for the "nominal case", after which a realistic application set-up is proposed and discussed in detail. For all simulations it is assumed that the "real plant" is described by the 164th order model (model A). The 84th (model B), 42nd (model C) as well as the two 5th order DAE and ODE models (models D and E) are used for the controller predictions. Note that when model A is used in the controller, there is no model/plant mismatch. As usual in distillation control, x_B and x_D are not controlled directly. Instead, an inferential control scheme that controls the deviations of the concentrations on trays 14 and 28 from the setpoints is used. Since for the standard operating conditions the turning point positions of the waves correspond approximately to these trays, one can expect good control performance with respect to x_B and x_D.

From Figure 3.61 it can be observed that even small changes in the inflow or feed conditions lead to significant changes in the wave positions, and thus of the concentrations on trays 14 and 28. However, the changes in the product concentrations x_B and x_D are in general quite small.

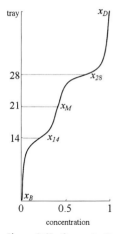

Figure 3.61 Concentration wave profile in the column.

3.3.3.3.1 Nominal Case

In this section the following assumptions are considered:

- That the full plant state is known. Since models A, B and C have the same differential states, the state feedback from model A (used as the plant) can be utilized directly when these models are employed in the NMPC controller. In the case of model D, the boiler, condenser and feed tray concentrations, as well as the positions of the wave inflection points, are needed. In order to obtain these five states, a least squares fit of the form stable wave profile to the 42 concentrations of model A is used.

- That the solution of the control problem is available immediately, which means that there is no delay between the measurement and implementation of control inputs.

- The feed concentration (considered the main source of disturbance) is measured. The terminal penalty matrix and terminal region are adjusted automatically during the calculations to the new steady state, which is an implicit function of the feed concentration.

According to the control considerations described above, only the concentration deviations from the setpoint on trays 14 and 28 are penalized in the cost-function [Eq. (3.236)]; that is, Q is semidefinite and only the diagonal elements 14 and 28 are nonzero and set to 1. The deviation of the control inputs from their steady-state values are not penalized – that is, $R = 0$.

The simulation results are presented in Table 3.7 and in the following figures. The chosen sampling time is $\delta = 30$ s, and all computations are carried out on a Compaq Alpha XP1000 workstation. For all simulations a disturbance scenario with the following step changes in the feed flow concentration (x_F) is used: at $t = 110$ s a step of -10%, at $t = 510$ s a step of $+15\%$, and at $t = 810$ s a step of -5%, back to the original steady state.

Table 3.7 Statistics of the necessary CPU times for one open-loop optimization problem for the QIH-NMPC approach using MUSCOD-II (time in seconds, using a Compaq Alpha XP1000 workstation) – Nominal case.

Model	$M = 5$ $(T_c = 150$ s$)$				$M = 10$ $(T_c = 300$ s$)$				$M = 20$ $(T_c = 600$ s$)$			
	Full iteration		1 iteration		Full iteration		1 iteration		Full iteration		1 iteration	
	Max.	Avge.	Max.	Avge.	Max.	Avge.	Max.	Avge.	Max.	Avge.	Max.	Avge.
A	22.19	3.91	2.01	0.85	40.35	4.27	3.89	1.41	60.92	6.31	7.71	2.97
B	*	*	*	*	*	*	*	*	*	*	*	*
C	0.93	0.45	0.24	0.20	2.06	0.82	0.51	0.37	6.89	2.00	1.30	0.92
D	0.34	0.11	0.07	0.05	2.48	0.26	0.14	0.09	4.79	0.55	0.34	0.24
E	*	*	*	*	*	*	*	*	*	*	*	*

* , simulations not performed.

, not feasible (i.e., maximum CPU time is greater than the sampling time).

The maximum and average CPU times necessary to solve one open-loop optimization problem for the QIH-NMPC scheme are shown in Table 3.7 for both the full iteration (iteration until a *KKT* tolerance: 10^{-6}) and the one-iteration approaches. It can be observed that the proposed QIH strategy, using the appropriate tool for optimization, is feasible for models C and D for all horizon lengths tried. In the case of the most complex model A, feasibility was achieved for a control horizon of $M = 5$. From the simulation results presented in the next few pages (Figs. 3.62–3.64) it can be observed that there is no significant difference between the control performances achieved with different control horizons. Additionally, the one-iteration approach leads to a similar control performance to the full iteration approach. In the figures, the quadratic objective function, Eq. (3.236), and the terminal penalty term (Mayer term), Eq. (3.241), are also presented to evaluate the feasibility of the optimization.

Due to the model plant mismatch, a slight steady-state offset can be observed for all models, but model A (in this case there is no model/plant mismatch because model A is also used as the plant). An interesting result can be seen in Figure 3.64 where, together with the results of the closed-loop controls, simulation results for the direct open-loop control are presented. In this case the final control action, after the control inputs have stabilized, and after rejecting the effects of the disturbance, is applied directly at the moment when the disturbance appears. An extremely slow response of the system can be observed for the direct open-loop control, whilst if the same control action is obtained in closed-loop after a certain sequence of command in only three to four sampling intervals, the system is stabilized very rapidly.

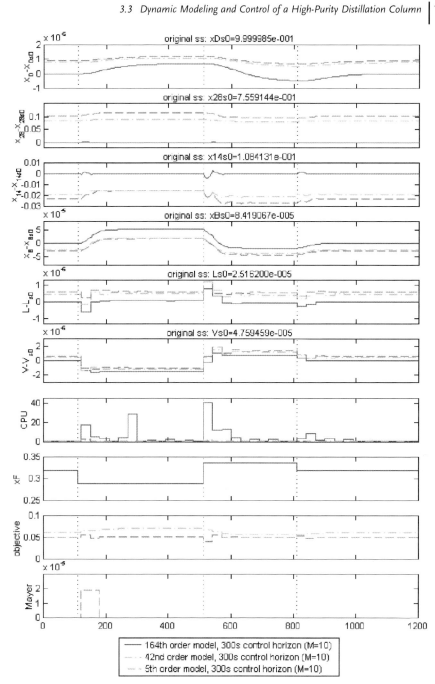

Figure 3.62 QIH-NMPC of the distillation column. Nominal case; comparison for different models, *M* = 10.

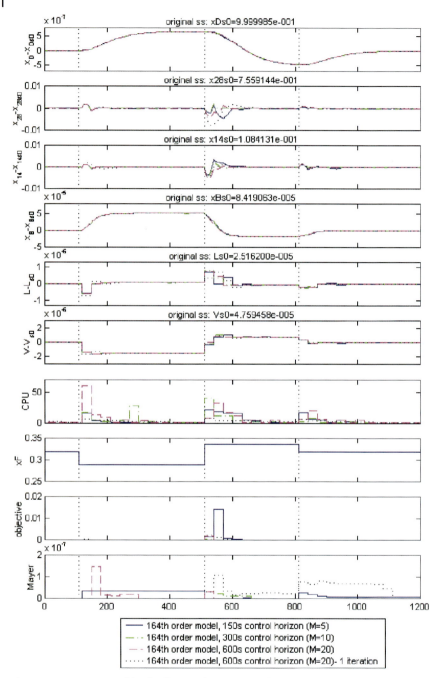

Figure 3.63 QIH-NMPC of the distillation column. Nominal case; comparison for different control horizons for the 164th order model and results with the one-iteration approach for $M = 20$.

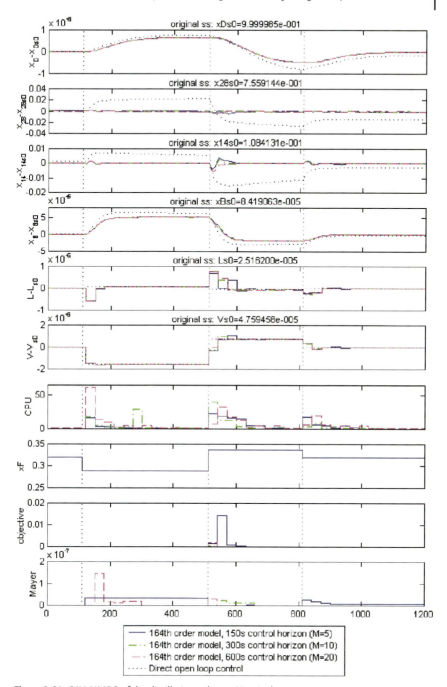

Figure 3.64 QIH-NMPC of the distillation column. Nominal case; comparison for different control

In order to demonstrate the advantage of an efficient NMPC approach such as QIH-NMPC, simulations with the zero terminal constraint NMPC scheme are performed for comparison. Usually, a much longer control horizon is necessary to allow feasibility of the zero terminal constraint. For the 5th order model (model D), a control horizon of $M = 20$ ($Tp = 600$ s) is sufficient, and the maximum CPU time increases to 6.8 s (average CPU time = 0.7 s). In the case of the 42nd order model (model C), a control horizon of $M = 50$ ($Tp = 1500$ s) is necessary [97]. The maximum CPU time increases to 131.5 s (average CPU time = 21.0 s). However, the optimization algorithm used shows excellent robustness, even in the case of rather large disturbances. The differences in control performance between the QIH and the zero terminal constraint NMPC are not significant.

3.3.3.3.2 Application Realistic Scenario

The challenge of a practitioner is to apply the theoretical results and to build an *implementable* NMPC controller, which means that a number of additional practical constraints must be confronted. The "ideal" controller setup described in the previous section cannot be implemented in practice, because of limitations in the availability of measurements and the computational time needed to elaborate the control action. In the following section all those specific tasks which make practical implementation of NMPC a challenging control problem are discussed, and the implemented solution in the proposed control setup is presented.

3.3.3.3.3 Estimation of the Unmeasured State Variables

In NMPC, current values of the state variables of the model are required to compute the predicted outputs. In practice, usually only a few (or even none) of these variables can be measured, and in the absence of total state-feedback it is necessary to implement a nonlinear observer to generate estimates of the unmeasured state variables using measured parameters. Nonlinear observer-based NMPC techniques are widely discussed in the literature [98–100].

In the application presented here, the Extended Kalman Filter (EKF) is used to estimate unmeasured parameters [101]. The EKF is appropriate for NMPC because it is able to accommodate to noise and modeling errors, it is computationally feasible in real-time, and it is conceptually straightforward. With the EKF, which is an extension to nonlinear systems of the linear Kalman filter, the basic idea is to perform linearization at each sampling time to approximate the nonlinear process as a time-varying system affine in the variables to be estimated. At each step of the time and measurement updates, a linear covariance matrix and the estimated states of the nonlinear system are updated, based on measurement functions. A brief description of the EKF is presented below; a detailed derivation of the EKF equations is available in the literature.

3.3.3.3.4 The Discrete EKF Algorithm

The discrete EKF addresses the general problem of trying to estimate the state $x \in \mathbb{R}^n$ of a discrete time-controlled process that is governed by the nonlinear stochastic difference equation:

$$x(k+1) = F(x(k), z(k), u(k), w(k)) \tag{3.269}$$

with a measurement $y \in \mathbb{R}^m$ that is

$$y(k) = h(x(k), z(k), v(k)) \tag{3.270}$$

The EKF linearizes the estimate around the mean and covariance, the discrete EKF equations being:

$$\hat{x}^-(k+1) = f(\hat{x}(k), \hat{z}(k), u(k), 0) \tag{3.271}$$

$$P_{k+1}^- = A_k P_k A_k^T + W_k A_k W_k^T \tag{3.272}$$

$$K_k = P_k^- H_k^T \left(H_k P_k^- H_k^T + V_k R_k V_k^T \right)^{-1} \tag{3.273}$$

$$\hat{x}(k) = \hat{x}^-(k) + K(y(k) - h(\hat{x}^-(k), \hat{z}^-(k), 0)) \tag{3.274}$$

$$P_k = (I - K_k H_k) P_k^- \tag{3.275}$$

where:

A is the Jacobian matrix of partial derivatives of $f(\cdot)$ with respect to x, that is

$$A_{[i,j]} = \frac{\partial f_{[i]}(\hat{x}(k), \hat{z}(k), u(k), 0)}{\partial x_{[j]}} \tag{3.276}$$

W is the Jacobian matrix of partial derivatives of $f(\cdot)$ with respect to w,

$$W_{[i,j]} = \frac{\partial f_{[i]}(\hat{x}(k), \hat{z}(k), u(k), 0)}{\partial w_{[j]}} \tag{3.277}$$

H is the Jacobian matrix of partial derivatives of $h(\cdot)$ with respect to x,

$$H_{[i,j]} = \frac{\partial h_{[i]}(\hat{x}(k), \hat{z}(k), 0)}{\partial x_{[j]}} \tag{3.278}$$

V is the Jacobian matrix of partial derivatives of $h(\cdot)$ with respect to v,

$$V_{[i,j]} = \frac{\partial h_{[i]}(\hat{x}(k), \hat{z}(k), 0)}{\partial v_{[j]}} \tag{3.279}$$

Equations (3.271) and (3.272) are the time update equations, while Eqs. (3.273), (3.274), and (3.275) are the measurement update equations. The first task during the measurement update is to compute the Kalman gain, K_k. The next step is actually to measure the process to obtain $y(k)$, and then to generate an *a posteriori* state estimate ($\hat{x}(k)$) by incorporating the measurement as in Eq. (3.274). The final step is to obtain an *a posteriori* error covariance estimate via Eq. (3.275). After each time and measurement update pair, the process is repeated with the previous *a posteriori* estimates used to project or predict the new a priori estimates ($\hat{x}^-(k)$).

The estimation is usually much faster than computation of the new control movement. Consequently, the estimation can be performed in parallel with the optimization but with a smaller sampling time, called the estimation sampling time (δ_e). In this way, at the next control sampling instance (δ_c), when the calculation of the new control movement begins, more recent information is available. Consequently, a better prediction – and therefore better control performance – can be achieved. The schematic representation of the NMPC with state observer is shown in Figure 3.65.

3.3.3.3.5 Estimation of the Disturbances

In the implemented NMPC approach, not only the states but also the unmeasurable disturbances (feed flow F and feed composition x_F), must be estimated. In order to estimate the disturbance, it is assumed that these parameters have unknown but constant values. Consequently, the disturbances can be considered as additional state parameters satisfying the differential equations:

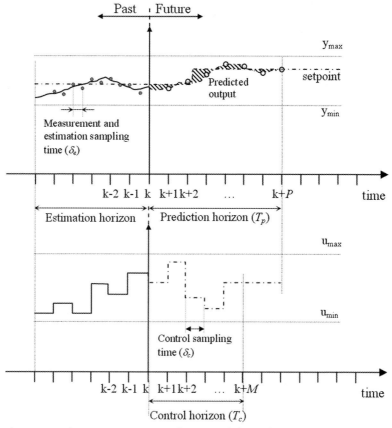

Figure 3.65 Schematic representation of NMPC with the implemented state estimation technique.

$$\frac{\mathrm{d}F}{\mathrm{d}t} = 0$$

$$\frac{\mathrm{d}x_F}{\mathrm{d}t} = 0 \tag{3.280}$$

3.3.3.3.6 Coping with Model/Plant Mismatch

In the practical implementation of NMPC, model/plant mismatch is a very common problem. From the simulation results obtained in the nominal case, it can be observed that in the case of model/plant mismatch the NMPC controller exhibits steady-state offset. Although, in the example presented here the main control objective – namely to keep the product concentration in a narrow high-purity band – is still met even in the case of differences between the model and plant, in a general NMPC implementation the model/plant mismatch must be taken into consideration. One general approach is to correct the setpoint values according to the deviations of the predictions from the measurements. In the application here, the model/plant mismatch is considered through the disturbance estimation, which leads to an implicit integral action of the controller. It has been shown in the simulations that it is possible to have a very similar steady-state concentration profile for all models for the same control inputs by allowing the two disturbance inputs (F and x_F) to have different values. Consequently, in the estimators, these two variables undertake the effect of the model/plant mismatch. In this way, an offset free control of the measured parameters is achieved.

3.3.3.3.7 Using Temperature Measurements instead of Concentration Measurements

In the case of the distillation column the states needed in the NMPC controller are concentrations for all stages (for models A, B and C) and wave positions (for models D and E). None of these parameters can be measured directly in practice. On the other hand, there is a unique relationship between the tray temperature and concentration (with the usually valid assumption of thermodynamic equilibrium), expressed by Eqs. (3.213) to (3.216). Thus, instead of controlling the concentrations, the corresponding temperature control leads to similar control performance. Consequently, only temperature measurements are considered, which can be performed easily and at relatively low cost. Controlling the deviations of the concentrations from their setpoints only on trays 14 and 28, leads to good control performance with regard to product purity. Hence, in the simulations only temperature measurements on trays 14 and 28 are considered, and all necessary parameters are estimated from these two measurements.

3.3.3.3.8 Delay Between the Computation and Implementation of the Control Inputs

In practice, the control inputs cannot be computed instantaneously. The effects of computational delay in the implementation of a controller have been investigated widely, and several model predictive control schemes taking into account this issue have been proposed for linear systems [102,103]. It has been shown that stability for the closed-loop system can be guaranteed if the delay is less than a certain threshold value. Here, the computational delay is taken into consideration by using a delay of one sampling time. That is, in every sampling time the first control input from the solution of the previous optimization problem is implemented, and simultaneously the computation of the new control movement is started using information up to the current sampling time. The solution of the new open-loop optimization problem is implemented at the next sampling time.

3.3.3.3.9 Real-Time Feasibility of the Optimization Problem

The major challenge in the practical implementation of NMPC is to assure a real-time feasibility of the solution of the nonlinear program. Because a nonlinear model is being used, the NMPC calculation usually involves a non-convex nonlinear program, which must be solved on-line at each time step. Efficient and reliable optimization approaches and tools are required in order to make the NMPC a real-time implementable technique. In the widely accepted implementation of the optimization problem, iterations are made until a certain tolerance is achieved (a $KKT = 10^{-6}$ is often used). However, in the case of the non-convex nonlinear program involved with the NMPC, no optimization technique has yet been identified which can guarantee the finding of a global optimum. Consequently, it cannot be assured that a certain tolerance will be always achieved in one sampling time period, even if the optimization is usually feasible. In addition, in order to preserve stability it is sufficient that at each time step there is a decrease in the finite horizon cost function, without necessarily finding the global minimum. Therefore, a suboptimal NMPC approach can be implemented by limiting the maximum number of iterations for optimization to a number that guarantees real-time feasibility. In this section it is shown that with the specific initial value embedding technique, implemented in MUSCOD-II, an excellent control performance can be achieved by performing only one SQP iteration at each sampling time. Thus, real-time feasibility is assured even for processes modeled by high-scale DAE systems and for long control horizons.

3.3.3.3.10 Controller Tuning

In the NMPC formulation described above, the main tuning parameters are derived from both the observer and the controller: these are the controller sampling time (δ_c), the estimation sampling time (δ_e), the control horizon (T_c), the prediction horizon (T_p), the weighting matrixes Q, R and Q_p from the controller side, and the weighting matrix and initial covariance matrix from the estimator side. The estimator parameters were chosen such that good and comparable estimation performance is achieved for all models. The same terminal penalty matrix Q_p, and weighting matrixes are used as in the nominal case ($R = 0$ and Q semidefinite with only the diagonal elements 14 and 28 equal to 1 and the rest set to 0). A control horizon $T_c = T_p$ was considered in all simulations, and results for different values are presented. The sampling time must be settled by a compromise between controller performance and on-line computation. Results with a sampling time of 30 s ($\delta_c = \delta_e$) are presented in Tables 3.7 and 3.8; however, it is clear that with the special one-iteration NMPC, a shorter sampling time can be used and hence the control performance can be improved.

3.3.3.3.11 Simulation Results

In order to estimate the computational burden and compare to the nominal case, similar simulations as in the nominal case were performed with the above-described NMPC scheme. The results are presented in Table 3.8. Similar computation characteristics can be observed as in the nominal case.

Table 3.8 Statistics of the necessary CPU times for one open-loop optimization problem for the QIH-NMPC approach using MUSCOD-II (time in seconds, using a Compaq Alpha XP1000 workstation) – with observer.

Model	$M = 5$ $(T_c = 150$ s$)$				$M = 10$ $(T_c = 300$ s$)$				$M = 20$ $(T_c = 600$ s$)$			
	Full iteration		1 iteration		Full iteration		1 iteration		Full iteration		1 iteration	
	Max.	Avge.	Max.	Avge.	Max.	Avge.	Max.	Avge.	Max.	Avge.	Max.	Avge.
A	9.77	2.77	1.57	0.87	38.46	5.89	3.92	2.17	71.56	13.49	6.05	3.35
B	4.64	0.66	0.68	0.46	19.03	1.41	1.53	0.94	44.68	9.98	2.34	1.97
C	1.12	0.32	0.34	0.23	3.74	1.16	0.75	0.59	8.49	3.53	1.80	1.25
D	0.62	0.14	0.09	0.08	0.96	0.27	0.23	0.14	2.73	0.71	0.47	0.37
E	0.14	0.05	0.06	0.03	0.29	0.10	0.11	0.07	0.87	0.30	0.28	0.20

Shaded area indicate that the process is not feasible (i.e., the maximum CPU time is greater than the sampling time).

By examining the computational times shown in Tables 3.7 and 3.8, it is clear that a shorter sampling time is feasible by using the one-iteration approach. In practical terms, there is no difference between the control performance with the one-iteration and full-iteration techniques. Furthermore, the use of a shorter estimation horizon can additionally augment control performance, as described above. These conclusions were observed for all five models, and are illustrated for model A in Figure 3.68. The results obtained in the nominal case for model A are also presented in Figure 3.69. Clearly, a better control performance can be observed in the nominal case due to the "ideal" assumptions considered. A comparison between the control performances achieved with the five models with the one-iteration controller using a control sampling time of $\delta_c = 10$ s and estimation horizon $\delta_e = 2$ s is shown in Figure 3.68. From this it can be observed that, if the estimator is correctly tuned, then very similar control performances can be achieved with all models, despite differences between them. The model differences

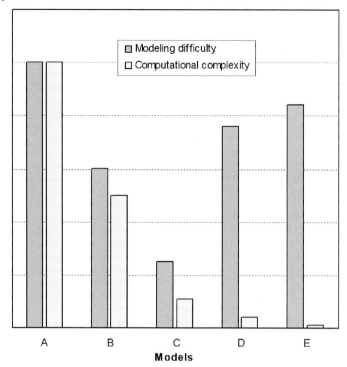

Figure 3.66 Comparison between modeling difficulty and computational complexity for the models.

are shown in Figure 3.67, where variation of the concentration profile for each model is shown compared to the plant, for the disturbance scenario used. The narrow bands indicate a good control performance, which maintains the concentration within tight boundaries. When model A is used in the controller there is no model/plant mismatch, and thus the bands approximately coincide. An important overlap of the concentration profiles can also be observed for model B, which means that this model provides a very good description of the real plant. The intersection points between the concentration bands for the plant and the models correspond to the measurement trays 14 and 28. By fixing these to positions of the wave profiles (by a tight control), the end concentrations are maintained in a high-purity boundary due to the stable form of the concentration profiles. The control performances obtained are similar for the control horizons used in the simulation ($M = 5, 10, 20$), but to demonstrate the performances of the optimization tool used the results for the computational most expensive case ($M = 20$) are presented. A similar control performance can be obtained independently of the model complexity, though the computational complexity decreases significantly for lower-order models. A relative comparison between computational complexity and modeling difficulty is shown in Figure 3.66. Models D and E are seen to be computationally very attractive, but the derivation of these reduced-order models is quite difficult.

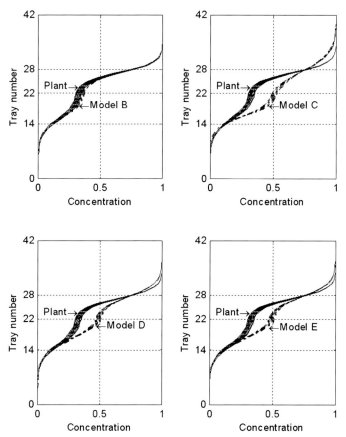

Figure 3.67 Concentration profiles for the plant and models during the control.

The use of very complicated deterministic models does not improve the control performance significantly, and they are computationally expensive and difficult to derive. When using the one-iteration approach, immediate feedback is practically possible for models D and E, even for a number of future control movement $M = 20$. For a shorter control horizon ($M = 5$), practical control implementation without delay is possible for models B, C, D, and E. The best choice seems to be model C, as this can be obtained with minimum modeling effort, and computational is not expensive. In fact, with model C immediate control feedback can be obtained, even for $M = 10$.

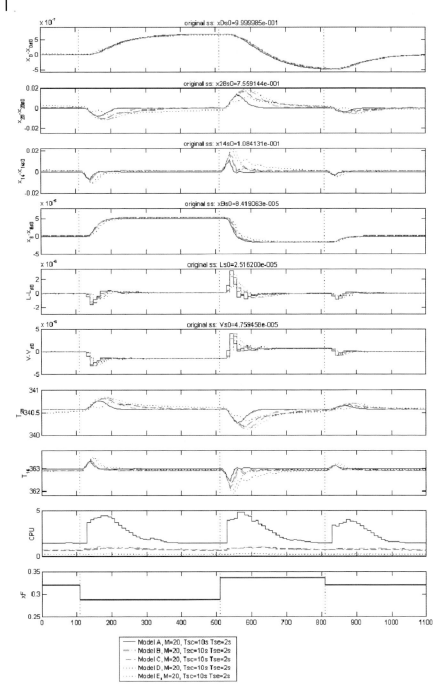

Figure 3.68 Control performances for the five models in the case of one-iteration QIH-NMPC approach.

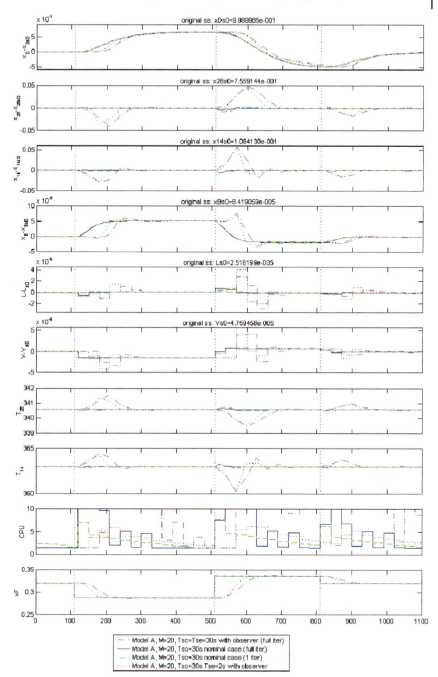

Figure 3.69 Comparison between controller performance for the nominal and realistic setup for model A.

3.3.4
Using Genetic Algorithm in Robust Optimization for NMPC of the Distillation Column

3.3.4.1 Motivation

In the NMPC controller, a constrained nonconvex nonlinear optimization problem must usually be solved on-line, as described previously. The major practical challenge associated with NMPC is that of computational complexity, which increases significantly with the complexity of the models used in the controller. It was shown in Section 3.3.3 that, by using the special, rapid on-line optimization algorithms which exploit the specific structure of optimization problems arising in NMPC, real-time applications can be feasible even for large-scale processes. However, the global solution of the optimization cannot be guaranteed. Additionally, despite the excellent performances of the specially tailored SQP optimization algorithm-based NMPC scheme (described previously), major disturbances can cause stability problems for the optimization algorithm. It is well-recognized that in industrial control applications, control performance and the robustness of the control scheme applied are of similar importance. Some methods for exploiting the advantages of GAs in order to improve the robustness of the NMPC schemes are presented in the following section.

3.3.4.2 GA-Based Robust Optimization for NMPC Schemes

A sequential approach was implemented in order to solve the on-line control problem imposed by the NMPC algorithms. This method uses different algorithms for the integration of differential equations and optimization. First, according to the algorithm of optimization, a sequence of control movements is considered; with this, the system of differential equations is numerically integrated to obtain the trajectory of the controlled variables. The scope function is then computed. The function of its value and the method of optimization yields a new sequence of control movements, after which the algorithm is repeated until the optimal sequence is obtained. Only the first value of the sequence is applied to the process. Here, it is proposed that the GA be used in combination with the special SQP algorithm presented previously in order to carry out optimization at each sampling time. In this respect, two approaches are presented to allow the advantages of both algorithms (GA and SQP) to be combined [104,105].

The first approach combines the two optimization methods into a single, more robust technique, in which GA is used for preoptimization to find the neighborhood of the global optima. Based on the best solution found after a certain number of generations, the SQP technique completes the optimization.

As the GA operators are designed to maximize the fitness function, in the preoptimization step the minimization problem described above must be transformed into a maximization problem. This can be done, for example, by utilizing the following transformation:

$$F = \frac{1}{1 + J(\cdot)} \tag{3.281}$$

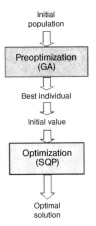

Figure 3.70 Sequential GA-SQP optimization algorithm.

The main idea of this approach is shown schematically in Figure 3.70. When this combination is used to solve the optimization in the NMPC algorithm, a very good robustness of the control scheme is achieved. However, due to preoptimization with the GA, the major advantage of the special SQP algorithm based on using information computed in the previous step, is lost.

The second approach (Fig. 3.71) uses SQP as the main optimization technique in the NMPC algorithm, but the control problem is also solved in parallel with GA. When a solution from the slower GA is available it is compared to that obtained with the SQP algorithm, and the better result is implemented. The main advantage of this approach is that it allows use of the previously described special SQP technique, without losing any of its performance. Only when the solution from the GA is implemented does the SQP need to be reinitialized. However, resetting the SQP algorithm in certain situations has proved to be very useful both for the control performance and especially for the robustness of the algorithm.

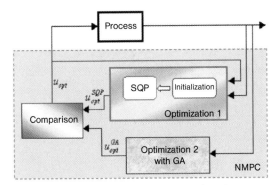

Figure 3.71 Parallel GA-SQP optimization algorithm.

For comparison, simulations were performed with the first principle model-based NMPC algorithm, using the aforementioned two different approaches in solving the on-line optimization problem. In addition, simulations with the classical algorithm, where the SQP is used only to solve the open-loop on-line optimization from the controller, are performed for comparison.

The 42nd order ODE model was used in the controller, and proved to be the best choice with regard to both computational complexity and modeling difficulty (see Section 3.3.3).

In the simulations, the following discrete form of the objective function was used:

$$J = \sum_{i=1}^{P} \left\| \begin{bmatrix} x_{14}(k+i) - x_{14}^{ss} \\ x_{28}(k+i) - x_{23}^{ss} \end{bmatrix} \right\|_Q^2 + \sum_{j=0}^{M-1} \|[u(k+j+1) - u(k+j)]\|_R^2 \qquad (3.282)$$

The results obtained with $Q = I$; $R = 0$; $P = 5$; $M = 1$; $T_{samp} = 30$ s, for a disturbance scenario with step changes in x_F and in the set points are presented in Figure 3.72.

The parameters of the GA are selected based on the characteristics of GA to quickly identify the neighborhood of the global optimum. The best values of fitness function found by the GA for one optimization in the case of different control horizon are presented in Figure 3.73. In all cases, and after only 20 generations, the GA found a close-to-optimum value, but then converged slowly to the exact value of the optimum. Consequently, in the sequential GA-SQP algorithm the GA was used

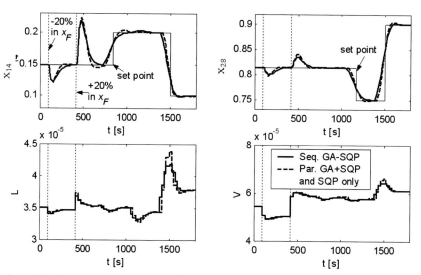

Figure 3.72 Comparative performance of three NMPC approaches for the following disturbance scenario: −20% step in x_F at $t = 90$ s; +20% step in x_F at $t = 420$ s; after that setpoint changes: $x_{14,sp} = 0.2$ at $t = 840$ s, $x_{14,sp} = 0.1$ at $t = 1500$ s, $x_{28,sp} = 0.75$ at $t = 1180$ s, $x_{28,sp} = 0.9$ at $t = 1500$ s.

initially only for 20 generations. The best solution found after 20 generations was close enough to the global optimum so that the SQP algorithm starting from this solution rapidly finds the exact value of the optimum, with minimal computational effort. Similar conclusions can be drawn from Figure 3.74, where the computational burden (expressed as the number of floating point operation, "flops") of the sequential GA-SQP algorithm is compared to situations when only the GA or SQP are used individually to solve optimization. Consequently, GA is seen to become computationally more attractive when the dimension of the optimization problem is increased, while the combined GA-SQP approach significantly reduces the computational burden in all cases. A population with 50 individuals for every optimization variable is used in all simulations.

The control performances obtained with the GA-based NMPC techniques are similar in this case, with the results obtained using only the SQP algorithm to solve the on-line control problem. Although the sequential GA-SQP-based NMPC improved control performance only slightly, its main advantage was the improved robustness of the on-line optimization (see Table 3.9) and the significant reduction in computational time needed to solve the open-loop control problem.

Table 3.9 Comparison of the robustness of the three optimization algorithms.

	SQP	Sequential GA-SQP	Parallel GA-SQP
No. of optimization failures in the case of 660 open-loop optimization with 20 disturbances	9	2	0

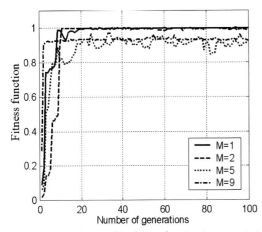

Figure 3.73 Evolution of the fitness function in one optimization with GA for different *M*-values.

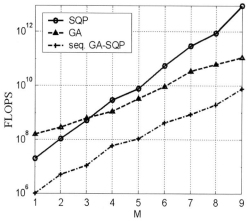

Figure 3.74 Comparison of the computational burden of the algorithms for different *M*-values. FLOPS = floating point operations.

For the parallel combined GA-SQP algorithm, both the control performance and computational burden are similar to those obtained with the SQP (the best results obtained when the algorithm does not fail), because in this case the control action implemented is usually that obtained from the SQP algorithm (which is faster). However, the parallel-running GA can significantly increase the robustness of the control structure, because whenever the main SQP fails, a close-to-optimum control action is always available from the GA.

3.3.5
LMPC of the High-Purity Distillation Column

Simulations with linear model predictive control (LMPC) were also performed for comparative purposes; the results obtained are shown in Figure 3.75.

The LMPC was seen to cope quite well with a similar disturbance scenario to that used for the NMPC (-20% step change in feed concentration at $t = 420$ s and $+20\%$ step change at $t = 6000$ s). However, both NMPC strategies clearly outperformed the LMPC in terms of both under/over-shooting and of the settling time. The best performance for the LMPC was achieved using the following tuning parameters: $T_p = 2$; $T_m = 1$; $Q = I$; $R = 0$. The same objective function was used as in the case of the NMPC algorithms.

3.3.6
A Comparison Between First Principles and Neural Network Model-Based NMPC of the Distillation Column

For most processes, the derivation of first principles models is an arduous, expensive, and time-consuming task, though as an alternative a nonlinear black-box model can be used. Their universal approximation property, together with their

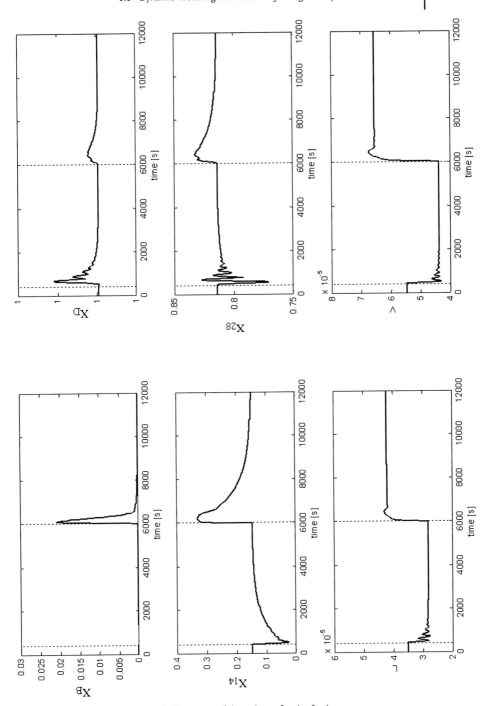

Figure 3.75 Simulation results with the LMPC of the column for the feed concentration disturbance: step −20% at $t = 420$ s, and +20% at $t = 6000$ s.

parsimony, makes ANN-based models attractive candidates for performing such a task. Many applications for both white-box and ANN model-based NMPCs have been reported in the literature, though as Henson remarked in his review [71], the NMPC literature lacks comparative studies between fundamental and empirical model-based techniques. Consequently, the main objective of this section is to compare the advantages and disadvantages of the first principle and ANN model-based NMPCs for a high-purity distillation column.

For all simulations it was assumed that the real plant was described by model A (a 164th order DAE). In the first principles model-based controller, models C and B respectively, were used and the differential states and model parameters F and x_F estimated from the two measurements (on trays 14 and 28) using a variant of the EKF.

For the comparative study, direct and inverse ANN models of the distillation column were derived in an input-output structure:

$$[\hat{x}_{14}(k+1),\ \hat{x}_{28}(k+1)] =$$
$$f_{ANN}\left(x_{14}(k-i)|_{i=\overline{0,2}},\ x_{28}(k-i)|_{i=\overline{0,2}},\ L(k-i)|_{i=\overline{0,2}},\ V(k-i)|_{i=\overline{0,2}};\ W\right) \tag{3.283}$$

$$[\hat{L}(k),\hat{V}(k)] =$$
$$f_{ANN}^{-1}\left(x_{14}^{sp}(k+1),\ x_{14}(k-i)|_{i=\overline{0,2}},\ x_{28}^{sp}(k+1),\ x_{28}(k-i)|_{i=\overline{0,2}},\ L(k-i)|_{i=\overline{1,2}},\ V(k-i)|_{i=\overline{1,2}};\ W_{inv}\right) \tag{3.284}$$

With these structures an adaptive model-based control scheme was implemented, as shown schematically in Figure 3.76.

According to this algorithm, the off-line trained ANN model is used to initialize the controller. Parallel with the control loop, a copy of the initial network is trained on-line. When the prediction error is less then 2%, the network from the control loop is replaced with the one adapted to the new conditions. Thus, the on-line training does not introduce any additional delay in the control movement compu-

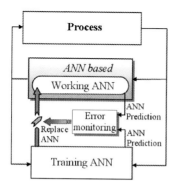

Figure 3.76 Schematic representation of the adaptive ANN-based control scheme.

tation. In order to obtain an ANN model with good generalization properties, a special training algorithm with Bayesian regularization is used [106]. The performances of the four controllers are compared for disturbance rejection and setpoint tracking in Figures 3.77 and 3.78, respectively. In all simulations, a sampling time of 30 s and prediction horizon equal to the control horizon ($P = M$) was considered. All of the controllers had similar control performances except for the inverse ANN model-based controller (invANNMPC). In the latter, there was a longer settling time, because the on-line open-loop optimization was eliminated and thus a small model/plant mismatch led to greater control errors. In addition, training the ANN in the inverse structure is usually more difficult because the pure derivative character of the inverted model can lead the process close to the limit of stability.

It must be emphasized that in this particular example the use of analytical models led to slightly better control performance. However, for complex processes in which first principle models can be derived only by applying broad simplifying assumptions, ANN models might be more accurate and lead to better control performance.

Subsequently, simulations for different control horizons (M) were performed in order to compare the computational burden of the control schemes. The CPU times needed to solve one open-loop optimization are presented in Table 3.10. ANN-based control techniques are very attractive from a computational point of view, due mainly to their simplicity, whereas first principles model-based NMPC algorithms are computationally very expensive. For these control schemes, special optimization tools must be used to reduce the computational time. The data listed in Table 3.10 show that the use of a multiple shooting technique (see Section 3.3.3) in MUSCOD leads to very good results being achieved.

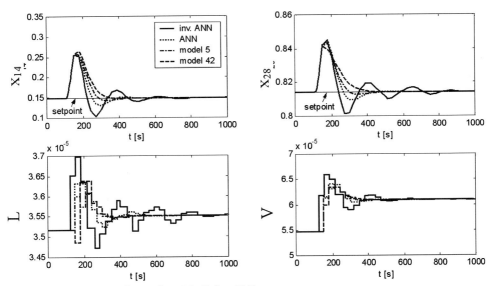

Figure 3.77 Performance of controllers ($M = 5$) for $+30\%$ step change in x_F at $t = 100$ s.

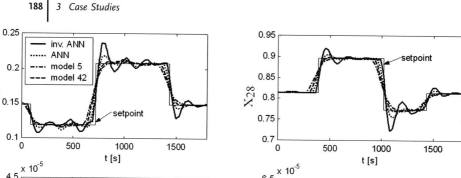

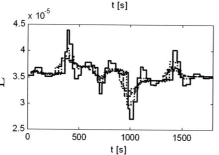 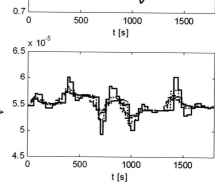

Figure 3.78 Performance of controllers ($M = 5$) for setpoint changes.

Table 3.10 CPU times per command (s), for different control horizon (M) in the case of the different controllers and software. Simulations were performed on a P-III/350 MHz computer.

	MATLAB						MUSCOD					
	$M = 5$		$M = 10$		$M = 20$		$M = 5$		$M = 10$		$M = 20$	
	Max.	Avge.	Max.	Avge.	Max.	Avge.	Max.	Avge.	Max.	Avge.	Max.	Avge.
5th ODE	310.2	189.5	516.1	380.3	*	*	1.4	0.3	2.9	0.6	5.9	1.5
42nd ODE	507.6	392.4	970.2	678.4	*	*	2.4	0.7	8.0	2.5	18.3	7.6
ANN	28.6	19.3	61.2	39.1	187.9	116.1	*	*	*	*	*	*
Inv. ANN	2.7	2.3	5.8	5.1	13.4	12.8	*	*	*	*	*	*

*, not performed.

A comparison of the main advantages and disadvantages of the ANN model-based control techniques and first principles model-based NMPC algorithms is provided in Table 3.11.

Table 3.11 Comparison between ANN and first principle model-based control techniques.

	ANN model-based control	First principle model-based control
Advantages	• General form; the same ANN algorithm will work for many different systems. • Accurate nonlinear models with very simple mathematical form, even for complex processes. • Can be trained to be relatively insensitive to noise. • Require little human expertise. • Computationally very attractive.	• Globally valid. • Allow understanding of the behavior of the system. • System theoretical results available for controller tuning and stability.
Disadvantages	• Crucial to define the right topology. • Require large amount of data to build the model. • Limited by the regions of data used to build the model. • Highly dependent on historical operations of the process. • Very few theoretical results are available concerning controller tuning and stability.	• Valid only for the particular systems, for which they were developed. • Require high-level human expertise and long development time. • For complex processes the necessary simplifying assumptions might harm the accuracy of the model. • Computationally very expensive. • State estimation is usually necessary, making controller tuning difficult.

3.3.7
Conclusions

In this section, a practically implementable NMPC approach was studied for the control of a distillation column, and the obstacles which must be overcome to make the NMPC applicable were outlined. It should be noted that some of the practical problems of NMPC (variable setpoint, implementation delay of the control action, early stopping of the optimization) can be solved with appropriate approaches and certain assumptions, without compromising the closed-loop stability guaranteed by the appropriate NMPC technique. However, in the case of model/plant mismatch and the use of observer, there are no general criteria to guarantee stability. Real-time feasibility and controller performance were discussed with regard to different model complexity. For this purpose, five different models have been developed and used in a QIH-NMPC controller. It was shown that using an appropriate optimization strategy (one SQP iteration with the special initial value embedding) and tool (MUSCOD-II), real-time feasibility can be achieved even for models with high complexity (164th order DAE) and control horizon with 20 future control actions. However, the complexity of the model used in the controller must be chosen based on a judicious compromise between modeling difficulty and controller performance on the one hand, and computational complexity on the other hand. In the case of the example presented here, it was shown that similar control performance could be obtained with models of different complexity. Moreover, if a well-tuned observer was implemented in an appropriate form, offset free and similar control performance was achieved even in the case of fairly important model/plant mismatch.

Additionally, two different NMPC approaches were proposed to improve the robustness of the NMPC controller, in which the advantageous properties of GAs in the successful solution of complex nonconvex nonlinear optimization problems are exploited. The first approach used GA for preoptimization in solving the on-line open-loop control problem. Starting from the best solution obtained, the SQP continued the solving; this approach improved both the control performance and the robustness of the NMPC. Additionally, the computational burden in this case was reduced. The second approach used GA to solve the optimization problem in parallel with the SQP, thereby improving the robustness of the NMPC, practically eliminating the controller failures due to failure of the optimization, and conferring on it great importance for practical NMPC implementations.

Finally, the first principles model-based control techniques were compared with the ANN model-based controllers. For this, a special adaptive ANN-based control structure was proposed and used with both the direct and inverse ANN models. By using ANN model-based controllers, it could be shown that very similar control performance can be achieved as in the case of white-box models, though the computational time required to solve one open-loop optimization problem in the controller was drastically reduced. The final conclusion drawn was that the use of fundamental models in the NMPC controller represents an attractive choice because they are globally valid and fewer process data are required for their development. However, when real-time feasibility of the optimization problem employed in the controller becomes an obstacle in practical implementation, special nonlinear black-box models, such as ANN, may be used instead without degrading the control performance, or specialized optimization techniques and/or tools (multiple shooting, MUSCOD) must be used.

3.3.8
Nomenclature

a_k	coefficient from equation for the liquid enthalpy of compound k as a function of temperature
A_k	coefficient from Antoine equation for compound k
b_k	coefficient from equation for the liquid enthalpy of compound k as a function of temperature
B	liquid molar flow from the reboiler (bottom product)
B_k	coefficient from Antoine equation for compound k
c_k	coefficient from equation for the liquid enthalpy of compound k as a function of temperature
C_k	coefficient from Antoine equation for compound k

D	liquid molar flow from the condenser (distillates)
E_s	terminal penalty term
F	molar flow of feed
h_j^L, h_j^V	total enthalpy of liquid and vapor phase, respectively on tray j, $j = M$ is the feed tray
$h_{k,j}^L, h_{k,j}^V$	enthalpy of pure compound k in liquid and vapor phase, respectively on tray j, $j = M$ is the feed tray
J_s	objective function for the NMPC controller
L_j	liquid molar flows from tray j, $j = 1,...,N$
L_M	liquid molar flows from feed tray
M	feed tray number
N	number of trays in the column
n_B	molar holdup in the reboiler
n_D	molar holdup in the condenser
n_j	molar holdup on tray j, $j = 1,...,N$
n_M	molar holdup of the feed tray
$p_{c,k}$	critical pressure of compound k
p_j	total pressure on tray j
$p_{k,j}$	vapor pressure of compound k on tray j
Q	weight matrix
Q_p	terminal penalty matrix
R	weight matrix
T_j	temperature on tray j
T_c	control horizon
$T_{c,k}$	critical temperature of compound k
T_p	prediction horizon
V_B	vapor molar flow from the reboiler
V_j	vapor molar flows from tray j, $j = 1,...,N$
V_M	vapor molar flow from the feed tray
x_B	mole fraction of the more volatile compound in the liquid phase in the reboiler
x_D	mole fraction of the more volatile compound in the liquid phase in the condenser
x_F	mole fraction of the more volatile compound in the liquid phase in the feed flow
x_j	mole fraction of the more volatile compound in the liquid phase on tray j, $j = 1,...,N$
y_B	mole fraction of the more volatile compound in the vapor phase in the reboiler
x_{z_0}, x_{z_N}	boundary concentration values for each section of the column for the wave model
y_D	mole fraction of the more volatile compound in the vapor phase in the condenser
y_j	mole fraction of the more volatile compound in the vapor phase on tray j, $j = 1,...,N$

Greek Symbols

α_r	relative volatility
β_k	coefficient from the enthalpy of vaporization for compound k
$\Delta h^v_{k,j}$	enthalpy of vaporization for compound k on tray j
Δp	pressure drop on each tray
μ_+, μ_-	asymptotic limits of the wave profiles
ϱ	shape parameter for the wave function
ξ	shape parameter for the wave function

3.4
Practical Implementation of NMPC for a Laboratory Azeotropic Distillation Column

3.4.1
Experimental Equipment

The experimental measures were obtained in an ILUDEST bubble cap tray column (Fig. 3.79); a schematic representation of the equipment is shown in Figure 3.80. This equipment has the following characteristics:
- 30 practical plates;
- operation volume 10 L;
- reboiler with quartz heating rod 2 kW;
- column head with solenoid controlled reflux-withdrawal divider and condenser of 0.2 m²;
- distillate cooler (cooling agent – water);
- feed heating system with quartz heating rod, capacity 0.5 kW;
- product receivers 5 L capacity each, to store the feed mixture, respectively to collect the bottom and head product;
- diaphragm pumps for feed and bottom product withdrawal;
- 39 sampling valves on every tray, for feed, bottom, distillate flows.

The data acquisition and control system provided with the system comprises the following components:
- Sensors: 18 temperature sensors places on every second tray, feed, reboiler and condenser; measurement of absolute pressure at top and differential pressure between the top and bottom, level probes, flow sensors, etc.
- Actuators in the plant: solenoid valves, heating elements (described above) liquid and vacuum pumps.
- Personal computer and accessories
- 19-inch "ILUDEST-MOS" unit as an interface between the distillation plant and the PC.

Experiments were conducted to determine the steady-state and dynamic characteristics of the column. An ethanol/water mixture was used in these experiments; this mixture forms an azeotropic mixture leading to a much more difficult modeling

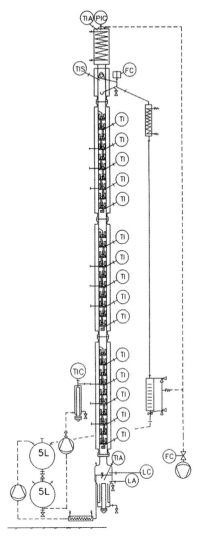

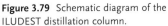

Figure 3.79 Schematic diagram of the ILUDEST distillation column.

Figure 3.80 The ILLUDEST distillation column from the Process Control Laboratory at UBB Cluj.

problem than for ideal mixtures. These difficulties and the modeling differences for this nonideal mixture is described in Section 3.4.3.

3.4.2
Description of the Developed Software Interface

The use of a computer-aided system is well suited to the control and regulation of distillation and rectification plants. The wide range of measurement and control tasks, the need for flexibility, ease of operation and the display of operating parameters means that using a computer with relevant software and peripherals is an ideal solution.

Although the above-described hardware has a software interface, the limited capabilities – together with a need to implement particular control schemes – made the development of a new software interface compulsory. During this development, the following issues were of particular attention:

- Actual communication with the distillation plant is accomplished through the ILUDEST-MOS unit on the first level. In this way, a hierarchical control structure with tasks distributed on two levels was implemented. This is necessary to ensure safe operation of the plant in case of computer failure. The PC is used as a means of communication between the operator and the electronics, for higher level controls, and for data storage.
- The most important criterion is: SAFETY OF THE OPERATION. For this, besides the hierarchical control structure, additional safety functions were implemented in the software.
- Simple operation through menu control.
- Flexible software architecture for easy future development.
- Implementation of some advanced model predictive control schemes.
- Possibility of remote access of the data via the Internet to develop a distillation "telelaboratory".

Communication between the PC and ILUDEST-MOS unit is established via the serial port of the computer using the RS-232 protocol, and a command protocol. The main window of the software with its main functions is presented in Figure 3.81.

The main window of the application provides a general view of the process. Current values of parameters are displayed in the schematic representation of the column. Here, the LEDs indicate correct operation of the column from a safety functions viewpoint. The history of the acquired data is presented in a chart, and a current temperature profile is also plotted. In this application, a remote communication kernel is also included which can be activated by the appropriate checkbox. If the "Enable Remote Server" checkbox is activated, the application will act as a server and allow remote access via the TCP/IP protocol from computers connected to the Internet. An access control check is implemented; thus, only authorized computers can establish data connection. The configuration window of the Data server is shown in Figure 3.82. In this window, the IP addresses of individual computers or Internet domains can be specified either to have access or deny to the Data Server.

The Main window of the client application is illustrated in Figure 3.83. The application provides remote control capabilities, which were tested successfully, for remote DAQ between the Laboratory of Process Control at Faculty of Chemistry and Chemical Engineering, from the "Babeş-Bolyai" University of Cluj, where the laboratory equipment is located, and the Automatic Control Laboratory from ETH, Zürich.

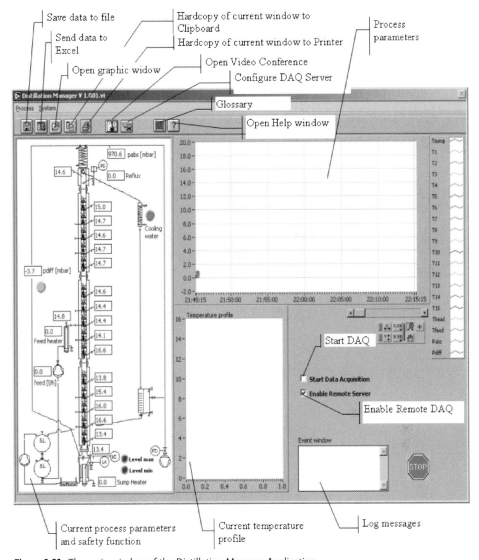

Figure 3.81 The main window of the Distillation Manager Application.

The application also provides a videoconference possibility. For this, from both the Server and Client applications the videoconference button launches Microsoft's Netmeeting application.

In controlling the distillation unit, the first criterion to take into consideration was *safe operation* of the system. For this, six safety functions were implemented in the software. These can be activated and configured in the Safety function window, which is opened from the System menu (Fig. 3.84). The column can be operated in either automatic or manual mode. In manual mode, the safety functions are not

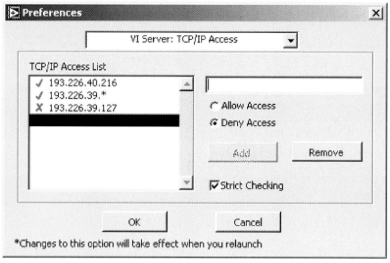

Figure 3.82 Configuration window for remote access.

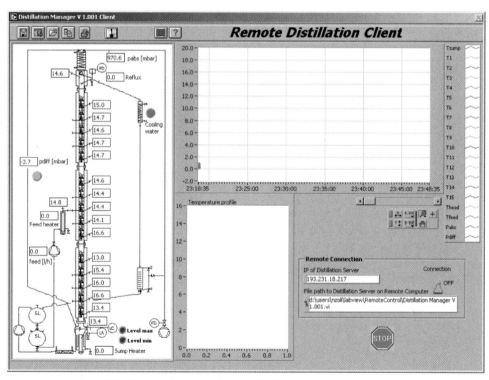

Figure 3.83 Main window of the Remote Distillation Client.

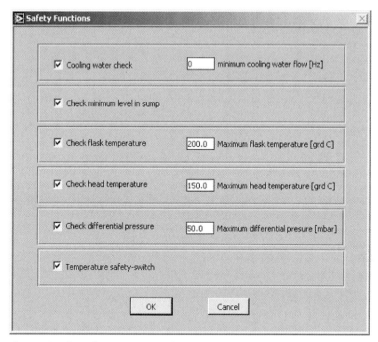

Figure 3.84 The Safety Functions window.

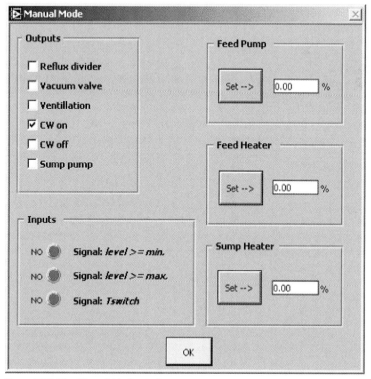

Figure 3.85 The Manual Mode window.

active, and process parameters may be set manually. The manual operation window (Fig. 3.85) is opened from the System menu, and safety functions are not active while this window is open.

The control loops of the distillation column can be configured via the Controller parameters window (Fig. 3.86), launched from the System menu. Four control loops were implemented in the system:

1. Sump level control: this is an on-off device – it indicates when the level has reached a maximum, but not the *actual* level. If the level reaches maximum, the sump pump is activated until the maximum level sensor is "off".
2. The feed temperature is controlled by a PID controller implemented with the classical discrete PID formula.
3. A control loop, which can control one of the column temperature (selected by the user) via the reflux ratio. Because of the particular features of the reflux system, implemented through a bipositional valve, the PID controller for this loop has not yet been implemented.

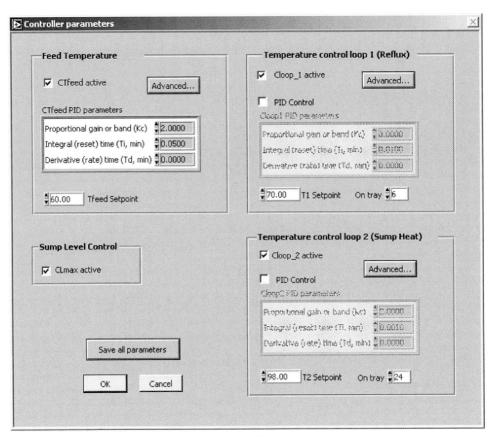

Figure 3.86 The configuration window of the controllers.

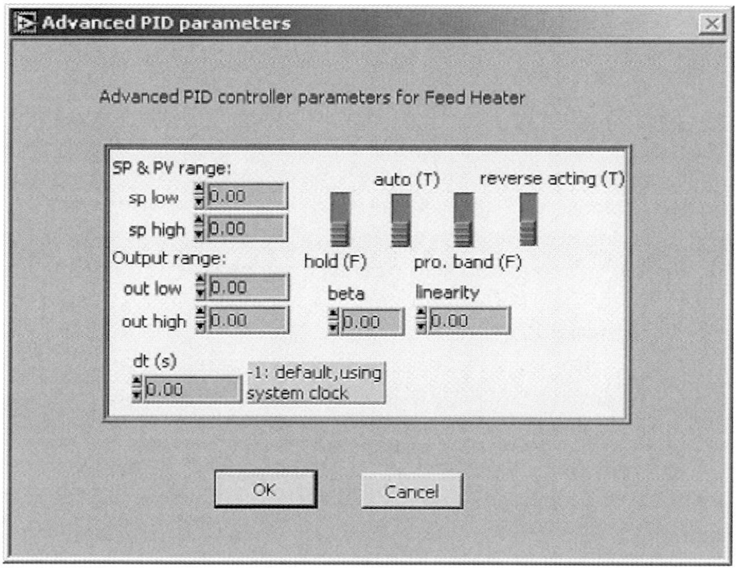

Figure 3.87 The Advanced Properties window for the PID controllers.

4. The second configurable control loop is for control of the second temperature value from the column (selected by the user) via the sump heater.

When the PID control checkbox is selected for loops 3 and 4, respectively, or for the second loop, the advanced button opens the window presented in Figure 3.87, where some advanced properties of the PID controller can be set by the user.

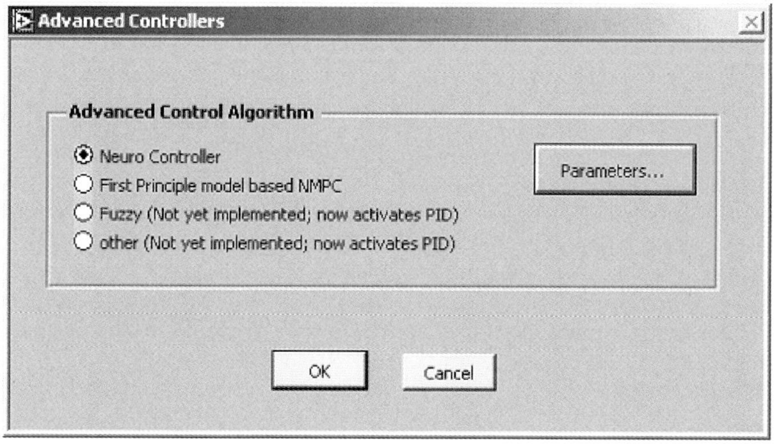

Figure 3.88 The configuration window of the controllers.

If the "PID control" checkboxes for the two configurable control loops are not checked, the advanced controllers can be activated for these parameters. In this case, the primary PID parameters are disabled and the "Advanced" button opens the "Advanced Controllers" window (Fig. 3.88). To date, two advanced controllers have been implemented: ANN model-based control, and first principle model-based control. One future objective of the project is the implementation of other advanced controllers (fuzzy, adaptive, etc.), the software permitting future developments.

Because the implemented controllers are for MIMO systems, the user can select the controller for either loop; the same control type will be selected automatically for the other loop. For the ANN-based control the "Advanced" button opens the ANN model selection window.

The advanced controllers are described in detail in the following sections. The controllers are implemented in MATLAB, and the control problem is solved, using the MATLAB GA toolbox developed by the author.

For this purpose, at every sampling time, the measurements are sent to MAT-LAB using the Dynamic Data Exchange (DDE) possibility in Windows environment. Here, the controllers using these data solve the control problem and send the obtained control actions back to the LabVIEW application; these are then sent to the process.

3.4.3
First Principles Model-Based Control of the Azeotropic Distillation Column

3.4.3.1 Experimental Validation of the First Principles Model
In a nonazeotropic distillation column, the vapor becomes steadily richer in the more volatile component as it passes through successive plates. In azeotropic mixtures, this steady increase in concentration does not take place, owing to the so-called "azeotropic points". For the mixture of ethyl alcohol and water used in these experiments, the concentration of the alcohol steadily increases until it reaches 95.6% by weight, when the composition of the vapor equals that of the liquid, and no further enrichment occurs (Fig. 3.89).

Computation of the equilibrium concentration for the azeotropic distillation, differs from that for ideal mixtures. Thus, model A described in Section 3.3.2.1 can be used in this case after replacing the relationships for the vapor–liquid equilibrium with the appropriate ones. Consequently, the general material and heat balances, Eqs. (3.203) to (3.212) and the general equations for computation of the enthalpies in the liquid and vapor phases, and pressure drops can also be used [Eqs. (3.216) to (3.222)], clearly using the appropriate parameters for the pure component of this mixture. In fact, the only equations which need to be modified from model A, are Eqs. (3.213) to (3.215).

The vapor–liquid equilibrium for the ethanol/water mixture can be computed with the following relationships:

$$y_i^e = \frac{x_i \cdot \gamma_i \cdot p_i^s}{p} \qquad i = 1, 2, \ldots, n \qquad (3.285)$$

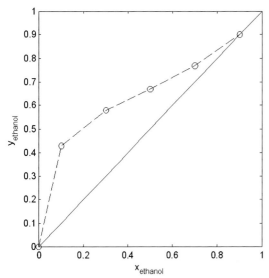

Figure 3.89 Vapor–liquid equilibrium curve for ethanol/
water system.

where the liquid nonideality in terms of the activity coefficients γ can be calculated
using the Wilson equation given below:

$$\ln(\gamma_i) = -\ln\left(\sum_{j=1}^{n} x_j \cdot \varphi_{ij}\right) + 1 - \sum_{k=1}^{n} \frac{x_k \cdot \varphi_{ki}}{\sum_{j=1}^{n} x_j \cdot \varphi_{kj}} \qquad i = 1, 2, \ldots, n \qquad (3.286)$$

Here, φ_{ij} is the interaction parameter between component i and component j given by:

$$\varphi_{ij} = \frac{V_j}{V_i} \cdot \exp\left(-\frac{\left(\lambda_{ij} - \lambda_{ii}\right)}{R \cdot T}\right) \qquad (3.287)$$

where V_i is the molar volume of pure liquid component i and λ_{ij} is the interaction
energy between component i and j. R the gas constant.

Adding to these equations the Antoine vapor pressure equation:

$$\log\left(\frac{p_i^s}{p}\right) = A_i - \frac{B_i}{T + C_i} \qquad i = 1, 2, \ldots, n \qquad (3.288)$$

$$\sum_{i=1}^{n} y_i^e = 1 \qquad (3.289)$$

the equilibrium concentration and temperature can be computed by solving a
system described by Eqs. (3.285) to (3.289) for each tray.

The parameters for the ethanol/water mixture are represented in Table 3.12.

Table 3.12 System parameters for ethanol/water binary azeotropic system.

Component	V_i	Antoine constants			Wilson constants	
		A_i	B_i	C_i	λ_{i1}	λ_{i2}
Ethanol	58.68	5.2314	1592.864	−46.816	1.0	288.9156
Water	18.07	5.1905	1730.630	−39.574	962.0073	1.0

Because the control parameter in this case is not directly the vapor flow from the reboiler, but the power of the sump heater (Q_w), this is introduced in the model by writing the energy balance for the reboiler:

$$0 = \eta_w \cdot Q_w - V_B \cdot h_B^V + L_1 \cdot h_1^L - B \cdot h_B^L - n_B \cdot \frac{dx_B}{dt} \cdot \left(h_{1,B}^L - h_{2,B}^L\right) \qquad (3.290)$$

From this equation, V_B can be obtained as a function of Q_w. Here, η_w is the reboiler efficiency.

Additionally, in order to cope with the nonideality of the system, the Murphy tray efficiencies are introduced. Two different efficiency parameters are used, one for the rectification section (η_r) and one for the stripping section (η_s), computed with the following equations:

$$\eta_r = \frac{V_j \cdot y_j - V_{j-1} \cdot y_{j-1}}{V_j \cdot y_j^e - V_{j-1} \cdot y_{j-1}}, \qquad j = a+1, \ldots, N \qquad (3.291)$$

$$\eta_s = \frac{V_j \cdot y_j - V_{j-1} \cdot y_{j-1}}{V_j \cdot y_j^e - V_{j-1} \cdot y_{j-1}}, \qquad j = 1, \ldots, a \qquad (3.292)$$

where a is the feed tray position, y_j^e is the equilibrium molar fraction in the vapor phase, obtained from system Eqs. (3.285) to (3.289), and y_j is the "real" molar fraction on tray j. Equations (3.291) and (3.292) are used to compute the real y_j, which is then used in the balance equations in the model.

The model validation is performed in two steps. First, the model is fitted to describe the steady-state behavior of the process. For this, the introduced efficiency parameters, η_w, η_r and η_s, are used as fitting parameters and four steady-state experiments were performed. The first two experiments were used to obtain the fitting parameters via minimization of the quadratic error criterion expressed by the difference between the measured (Tm_{kj}) and modeled (T_{kj}) tray temperatures:

$$J_{ss} = \sum_{k=1}^{N_{ex}} \sum_{j=1}^{N_m} \left(Tm_{kj} - T_{kj}\right)^2 \qquad (3.293)$$

where N_{ex} represents the number of experiments, and N_m the number of temperature measurement points in the experiments. Thus, the optimization problem used to fit the model can be described by:

$$\min_{\eta_w, \eta_r, \eta_s} (J_{ss}) \tag{3.294}$$

The results obtained by solving this optimization problem were as follows: $\eta_w = 0.875$; $\eta_r = 0.786$; and $\eta_s = 0.731$.

The value of the performance criterion [Eq. (3.293)], for the data sets used to fit and test the model, and the experimental conditions are presented in Table 3.13.

Table 3.13 Experimental data and model validation results for the steady-state modeling.

	Experiment	F [L h^{-1}]	x_F	T_F [°C]	R [s/s]	Q_v [%]	J_{ss}
Used to fit the model	SS1	2	30.26	58.7	5/1	50	2.4
	SS2	2	30.26	58.7	10/1	40	
Testing data	SS3	2	30.26	58.7	7/1	60	4.6
	SS4	2	30.26	58.7	10/1	50	3.7

The tuned results obtained for the parameters of η_w, η_r and η_s obtained above are illustrated in Figure 3.90 for experiment SS1, and in Figure 3.91 for the testing experimental data SS3.

A fairly good steady-state modeling capability can be observed from Figures 3.90 and 3.91. In the second step, experiments to identify the dynamic behavior of the system have been performed. The experimental scenario was as follows: from the initial steady state, a step change between 40% and 60% at $t = 0.2$ h, of the sump heater power was used. When the system had arrived close to the new steady state, a backward step change of the sump heater power from 60% to 40% was applied at $t = 1.05$ h. The experimental data were then divided in two parts. The first part (experiment D1) was used to fit the model, and the second part (experiment D2) to test to model.

The liquid holdups (in mL) in the reboiler (n_B), on the trays (n_j, $j = 1, ..., 30$), and condenser-tray (n_D) were used to tune the model. In this case, the sum-squared error of the temperature obtained from the model compared to the measurements at different time instances was used as the performance criteria:

$$J_{dyn} = \sum_{k=1}^{T_f} \sum_{j=1}^{N_m} \left(Tm_{kj} - T_{kj} \right)^2 \tag{3.295}$$

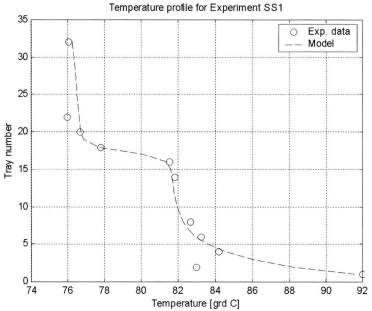

Figure 3.90 Comparison between model and experiment for data SS1 used to fit the model.

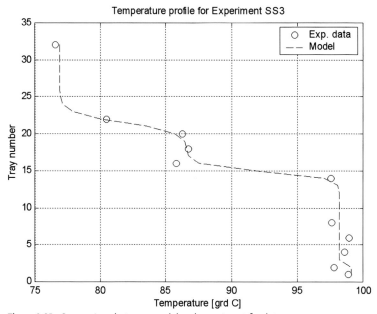

Figure 3.91 Comparison between model and experiment for data SS3 (test data).

where T_f is final time of the experiment. Temperature values with a step of 1 min were used in the optimization problem expressed below:

$$\min_{n_B, n_j, n_D} \left(J_{dyn} \right) \tag{3.296}$$

The model parameters obtained are presented in Table 3.14. The results for selected trays (reboiler, trays 3, 6, 19, and head) are presented in Figure 3.92.

Table 3.14 Experimental data and model validation results for the dynamic model identification.

Exp.	Q_w	R [s/s]	F [L h^{-1}]	x_F	T_F, [°C]	Initial values used in SS1-SS4 [mL]			Final values, after optimization [mL]			J_{dyn}
						n_B	n_j	n_D	n_B	n_j	n_D	
D1	40% → 60%	5/1	2	27.5	56.4	2430	10	10	2106.4	8.1	13.7	168.9
D2	60% → 40%	5/1	2	27.5	56.4	2430	10	10	2106.4	8.1	13.7	2117.4

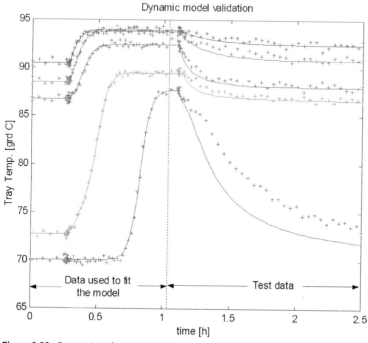

Figure 3.92 Comparison between experimental data and dynamic model (+, experimental data; solid lines represent the data obtained from the model).

It can be observed from this figure, that the model describes fairly well the dynamic characteristics of the process. However, it should be emphasized that for an appropriate dynamic model, many more experiments must be conducted in order to determine the parameters and to validate the model. For example, a study of the effects of feed composition and flow would be necessary. The more comprehensive dynamic model identification represents one of our future research objectives.

3.4.3.2 First Principle Model-Based NMPC of the System

During the past few years, several research groups have approached the model predictive control of a distillation column, either continuous or discontinuous. Thus, Fileti, Cruz and Pereira (2000) analyzed and compared the performances of several control strategies applied to batch distillation [107]. Later, Hapoglu, Koca and Alpbaz or Karacan applied different techniques of MPC to control the overhead temperature [108–110]. Diehl et al. [111] presented experimental results from the implementation of the efficient NMPC approach described in Chapter 2. For the FPNMPC of the distillation column, the control inputs considered in the present case study were the sump heat power (Q_w) and the reflux rate (R). The controlled parameters were the temperatures on trays 3 and 25. The control algorithms described in Chapter 2, Section 2.3.3, cannot be directly implemented to the above-described laboratory distillation plant, due to the particular nature of the reflux valve, which is bi-positional. When the valve is open, the entire liquid flow from the condenser is extracted from the system as the distillate; however, when the valve is closed, the liquid is reintroduced into the column as the reflux flow. Consequently, this device is digital, and does not allow the implementation of any analogue value for the reflux flow rate given by the algorithms described so far.

In this section, hybrid control architecture is suggested for the pilot distillation plant. According to this control algorithm, one control input (Q_w) is an analogue input which can take any value in the interval of 0 to 100%. The second control input, instead of an analogue value for R (which is not applicable in this system), is developed as follows.

Within the sampling interval T_s, a smaller sampling interval δ_s is considered, such that $N\delta_s = T_s$, with $N \in \mathbb{N}$. The digital control input will be a vector of the form:

$$\mathbf{u} = [u_1, \, u_2, \dots, u_N] \tag{3.297}$$

with

$$u_j\big|_{j=\overline{1,N}} = \begin{cases} 0 & \text{if the valve is closed (reflux)} \\ 1 & \text{if the valve is open (distillate)} \end{cases} \tag{3.298}$$

The control value $u_j = 0$ means that the reflux valve is closed for a period of δ_s between moments $(j-1) \cdot \delta_s \leq \tau < j \cdot \delta_s$ from the sampling period of T_s. The schematic representation of the control inputs for the proposed hybrid system is depicted in Figure 3.93. The optimization problem from the controller, that must be solved on-line in each sampling period T_s is as follows:

$$J(Q_w, \mathbf{u}) = \sum_{k=1}^{P} \left[\lambda_1 \cdot \left(T_3(k) - T_{3,s}(k) \right) + \lambda_2 \cdot \left(T_{25}(k) - T_{25,s}(k) \right) \right]^2 \qquad (3.299)$$

$$\min_{Q_w \in \mathbb{R}^q; \, u_j|_{j=\overline{1,N}} \in \{0,1\}} \left(J(Q_w, \mathbf{u}) \right) \qquad (3.300)$$

where $\lambda_{1,2}$ are weight coefficients, $T_{3,s}$ and $T_{25,s}$ are the setpoints for the temperatures on trays 3 and 25, respectively, and P is the prediction horizon, $\mathbb{R}^q$ is the subdomain of real values from the interval $[Q_{w,min}, Q_{w,max}]$, and $u_j|_{j=\overline{1,N}} \in \{0, 1\}$ are the binary values of the control inputs in the interval T_s.

The optimization problem described by Eqs. (3.299) and (3.300) represents a special mixed real and integer optimization problem. The parameters to obtain from the solution of this problem are formed from one real value Q_w and N binary integer values u_j.

The solution of such optimization problems represents a major numeric challenge. The special structure of the problem can be exploited very well if a genetic algorithm (GA) is used to solve the optimization. The feature to use binary representation of optimization parameters to solve the optimization makes this an alluring tool for the problem. Thus, for the GA we may consider only two parameters to be obtained from the optimization. The first, is the real Q_w value, which can be coded on 16 bits to have the necessary precision in the real interval [0 100], from which this control input may take values. The second optimization state, $\mathbf{u}$, can be considered as a binary value from the interval $\mathbf{u} \in [\underbrace{000 \ldots 0}_{N}, \ldots, \underbrace{111 \ldots 1}_{N}]$.

The obtained binary value for the Q_w is decoded in the real representation to be implemented in the process, while the binary $\mathbf{u}$ is interpreted in binary form. Every bit from $\mathbf{u}$ corresponds to the appropriate reflux valve position (0 = *closed*; 1 = *open*).

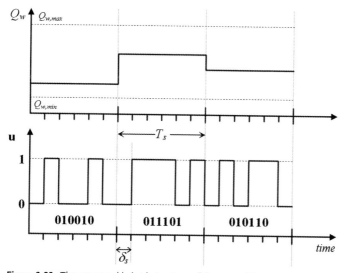

Figure 3.93 The proposed hybrid structure of the control inputs.

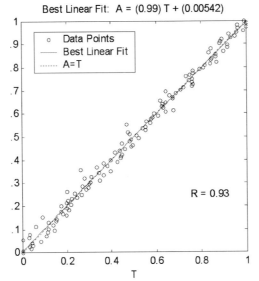

Figure 3.96 ANN prediction for the training data for T_{25}.

and $b = 0$, the ideal prediction is $A = T$. In our case, it can be observed that although the a and b values are close to their ideal values and the correlation coefficient (R) is also high, the prediction is not ideal. These results suggest uniformly distributed errors, which most likely are due to measurement errors. Thus, one should expect that the network would filter the noise and give fairly good prediction.

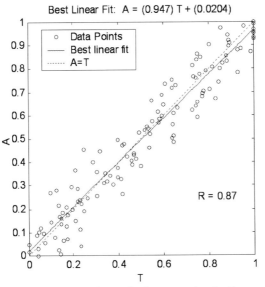

Figure 3.97 ANN prediction for the testing data for T_3.

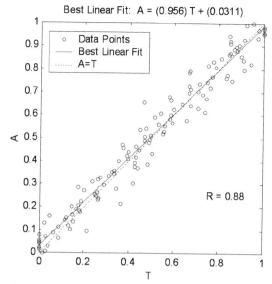

Figure 3.98 ANN prediction for the testing data for T_{25}.

The ANN model-based NMPC was implemented in the software interface. The optimization (with GA) and the control movement computation is performed in MATLAB. By introducing the ANN model, presented above, a good control performance is achieved. The ANNMPC was tested for setpoint tracking; for this, a setpoint change was applied to the system for both controlled parameter (T_3 and T_{25}) but at different moments. The results are shown in Figure 3.99.

The controller parameters $\lambda_1 = \lambda_2 = 1$, $P = 20$ (600 s), and $M = 1$ were used. The successive-recursive prediction algorithm, described in Chapter 2 (Section 2.3.5.4) is used to perform the $P > 1$ prediction.

The parameters of the GA used in the optimization are: 50 members in the population; cross-over probability = 1; mutation probability = 0.03; limits for the inputs $Q_{pv} \in$ [0 100], $\mathbf{u} \in$ [000000 111111]; codification bits = [16 6]; maximum number of generations = 100; With these parameters, an open-loop optimization was solved in less then 20 s, which is less than the sampling time used, $T_s = 30$ s. Consequently, the ANMPC controller in the above-described structure is feasible for real-time implementation.

3.4.5
Conclusions

In this section the practical implementation of two nonlinear model-based predictive controller to a nonideal azeotropic distillation column was presented. First, the 30-tray ILLUDEST pilot laboratory equipment was described, after which the software interface developed by the present author was presented in detail. This

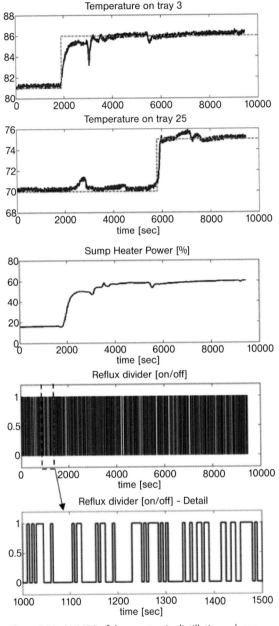

Figure 3.99 ANMPC of the azeotropic distillation column.

interface joins the two most important characteristics of computer interfaces: flexibility and safety operation. Additionally, the software provides the possibility of remote experiments from computers connected to the Internet.

For control of the system, a two-inputs–two-outputs control architecture was considered. The control inputs were the sump heater power and reflux rate, respectively, while the controlled outputs consisted of temperatures on trays 3 and 25, respectively. The sump heater used was an analogue device, while the reflux rate was set via a digital on/off reflux divider valve. This led to a mixed binary integer-real structure. Because of these particular features of the equipment, the control of the system is a challenging task. A novel hybrid control approach was proposed which, in combination with GA, exploited the special structure of the control system.

Two models of the distillation column have been developed for use in the NMPC control scheme. The first model was analytical in nature, and based on a detailed first principle modeling approach. Experiments were performed to fit the model for both the steady-state and dynamic behavior of the system. This model was used for prediction in the proposed NMPC. It was shown that although the dynamic model described the process dynamics fairly well, the first principle model-based NMPC approach was not feasible for real-time implementation. The solution of one open-loop control problem took about 2 h on a P-III/600 MHz computer. Consequently, the use of faster software (i.e., MUSCOD) instead of MATLAB or the use of a faster model for prediction is necessary. Here, the latter approach was adopted. In the previous section, the advantages of ANN model-based control compared to the first-principles model-based algorithms were emphasized. These consist of much faster computation and lower modeling difficulties. These are especially true for the azeotropic distillation where, because of the nonideal mixture, the process models are even more complicated. Thus, the necessary computational burden is much higher and the development of accurate models for these systems is an arduous task.

An ANN model adapted to the special structure of the mixed logical dynamic system has been developed and used in the novel GA-hybrid control system. This algorithm was implemented in the distillation system, achieving very good control performance.

3.5
Model Predictive Control of the Fluid Catalytic Cracking Unit

3.5.1
Introduction

For over 60 years, catalytic cracking has been one of the main processes in petroleum refining, having passed through spectacular development [112]. The fluid catalytic cracking unit (FCCU) has, during the past few decades, become the "test bench" of many advanced control methods. Today, both academia and industry are expressing great interest in the development of new control algorithms and in their efficient industrial FCC implementation, as successful results are usually of major economic benefit [113]. The catalytic cracking process is complex both from the modeling and from the control points of view [114–117].

The dynamic mathematical model development implies some assumptions, taking into account the specific aspects of the process. The complex nature of the feed oil assumes a lumped kinetic mechanism for the treatment of the cracking process. Both reactor and regenerator mass and heat transfer are complex. An adiabatic plug flow reactor model is usually used for the riser. Two zones frequently describe the regenerator model: (i) a dense bed zone (with dense phase as a CSTR model, but gaseous phase as a plug flow reactor model); and (ii) an entrained catalyst zone (plug flow model) [118].

The control system design and implementation must solve an array of challenging tasks, the main problems being the multivariable character of the process, presenting strong interactions, the nonlinear behavior leading to a need for nonlinear control, and a demand to operate the unit in the presence of material and operating constraints. Additionally, the control system must cope with both long and short time constants, as well as facing changing operating conditions, in the presence of unmeasured disturbances [119]. As a consequence, Model Predictive Control (MPC) has proved to be a good candidate for the advanced control of FCCU, due to its multivariable structure, direct constraints handling, and economic optimization characteristics [120–125].

Based on these preliminary aspects, this section presents the development of a mathematical model for a UOP (Universal Oil Products)-type FCCU and the associated dynamic simulator. Different MPC schemes are investigated and tested by dynamic simulation, revealing interesting aspects from the perspective of industrial implementation.[1]

3.5.2
Dynamic Model of the UOP FCCU

The FCCU, for which the mathematical model has been developed and the Model Predictive Control study then performed, is presented in Figure 3.100.

The mathematical model developed for the UOP-type FCCU is based on the mechanistic Amoco Model IV FCCU [118]. Compared with Amoco Model IV, the new mathematical model describes a different FCCU type, both from operation and construction points of view. The main model characteristics are related to the following aspects:

- different geometric dimensions and relative position define the reactor and regenerator, compared with the Model IV case;
- the reactor model uses a Weekman kinetic scheme [126] to describe the cracking process;
- the regenerator of the UOP FCCU operates in partial combustion mode;
- catalyst circulation is described, including spent and catalyst valves on catalyst circulation lines. These valves are used as main manipulated variables for FCCU control.

---

1) Parts of the material presented in this section are reprinted from Cristea, M. V., Agachi, S. P., and Marinoiu, V., Simulation and model predictive control of a UOP fluid catalytic cracking unit. *Chemical Engineering and Processing*, *42*, 67–91. Copyright 2002, with permission from Elsevier.

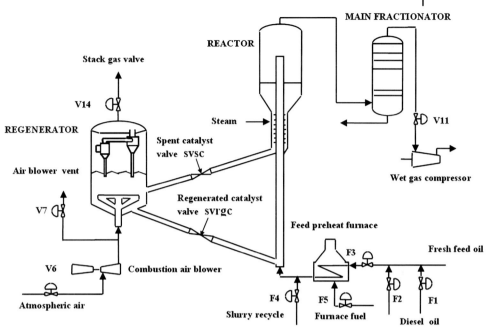

Figure 3.100 Schematic representation of a UOP-type fluid catalytic cracking unit (FCCU).

The FCCU dynamic model has been developed on the basis of reference construction and operation data from an industrial unit. The described model is rather complex, and has succeeded in capturing the major dynamic behavior of UOP-type FCCU [127,128]. The model includes the main reactor-regenerator subsystems: feed and preheat system, reactor, regenerator, air blower, wet gas compressor and catalyst circulation lines.

The main aspects of the new model are outlined in the following sections.

3.5.2.1 Reactor Model

The development of a new mathematical model for the reactor necessitated a thorough survey, selection and then synthesis, based on a large variety of models presented in the literature. The three-lump model has been considered as adequate for the global description of the phenomena that take place in the reactor. The reactor is divided into two parts: the riser and the stripper. The riser model is built on the following assumptions: ideal plug flow and very short transient time (the residence time in the riser is very short compared to other time constants, especially with the regenerator time constants [112,118,126,127]). It is modeled by mass balance, describing the gasoline and coke + gases production based on Weekman's triangular kinetic model [126]. The mixed nonlinear differential and algebraic system of equations also accounts for the amount of coke deposited on the catalyst and for the cracking temperature dynamics [129]. The reactor is illustrated in Figure 3.101.

A detailed description of the reactor model is presented in the following sections.

3.5.2.1.1 Mass Balance for the Riser

The mass balance for the feed is described by the equation:

$$\frac{dy_f}{dz_a} = -K_1 y_f^2 [COR] \Phi t_c \tag{3.301}$$

The mass balance for the gasoline is described by the equation:

$$\frac{dy_g}{dz_a} = (\alpha_2 K_1 y_f^2 - K_3 y_g)[COR] \Phi t_c \tag{3.302}$$

where:

$$K_1(\theta) = k_{r1} e^{\frac{-E_f}{RT_0(1+\theta)}} \tag{3.303}$$

$$K_3(\theta) = k_{r3} e^{\frac{-E_g}{RT_0(1+\theta)}}, \quad \theta = (T - T_0)/T \tag{3.304}$$

$$\Phi = \phi_0 e^{(-\alpha\, t_c [COR] z_a)} \tag{3.305}$$

$$\phi_0 = 1 - m^* C_{rgc} \tag{3.306}$$

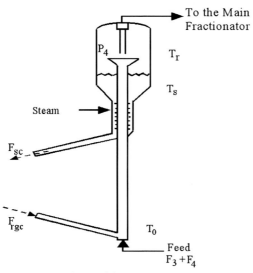

Figure 3.101 Scheme of the FCCU reactor.

The inlet temperature in the riser, T_0, is determined by the heat balance equation [114]:

$$T_0 = \frac{F_{rgc} C_{pc} T_{reg} + F_f C_{pf} T_2 - \Delta H_{evp} F_f}{F_{rgc} C_{pc} + F_f C_{pfv}} \qquad (3.307)$$

The term $K_1 y_f^2 [COR]$ represents the kinetics of the feed, and $K_3 y_g [COR]$ the kinetics of the gasoline; Φ is a function of catalyst deactivation due to coke deposition; ϕ_0 the reduction of catalyst activity due to the coke resident on the catalyst after regeneration; t_c residence time in the riser; $\alpha_2 = k_1/k_2$ fraction of feed oil that cracks to gasoline. This model develops the models presented by Lee and Groves [130], Shah et al. [131], and Hovd and Skogestad [129]. The amount of coke produced is described by the following correlation taken from Voorhies and Kurihara [132]:

$$C_{cat} = K_c \sqrt{\frac{t_c}{C_{rgc}^N}} e^{\frac{-E_{cf}}{RT_r}} \qquad (3.308)$$

The fraction of coke on the spent catalyst leaving the riser is:

$$C_{sc1} = C_{rgc} + C_{cat} \qquad (3.309)$$

The constant values m^* and N have been used to perform a good fit of the mathematical model with operating data from the industrial unit.

3.5.2.1.2 Heat Balance for the Riser

$$\frac{d\theta}{dz_a} = \frac{\Delta H_f F_f}{T_0 (F_{sc} C_{pc} + F_f C_{pf} + \lambda F_f C_{pd})} \frac{dy_f}{dz_a} \qquad (3.310)$$

The amount of gases produced by cracking is described by the equation:

$$F_{wg} = (F_3 + F_4)[C_1 + C_2 (T_r - T_{ref})] \qquad (3.311)$$

Constants C_1 and C_2 have been fitted based on data from the industrial unit.

The stripper model is of CSTR type (mass and heat balance), evaluating the temperature in the stripper and the fraction of coke on spent catalyst.

3.5.2.1.3 Mass and Heat Balance for the Stripper

$$\frac{dT_s}{dt} = \frac{F_{rgc}}{W_r} (T_r - T_s) \qquad (3.312)$$

$$\frac{dC_{sc}}{dt} = \left[F_{rgc} (C_{rgc} + C_{cat}) - F_{sc} \cdot C_{sc} - C_{sc} \frac{dW_r}{dt} \right] \frac{1}{W_r} \qquad (3.313)$$

$$\frac{dW_r}{dt} = F_{rgc} - F_{sc} \qquad (3.314)$$

3.5.2.1.4 Pressure Balance for Riser Bottom Pressure Determination

$$P_{rb} = P_4 + \frac{\varrho_{ris} h_{ris}}{144} \qquad (3.315)$$

$$\varrho_{ris} = \frac{F_3 + F_4 + F_{rgc}}{v_{ris}} \qquad (3.316)$$

$$v_{ris} = \frac{F_3 + F_4}{\varrho_v} + \frac{F_{rgc}}{\varrho_{part}} \qquad (3.317)$$

The amount of catalyst in the riser is determined by the equation

$$W_{ris} = \frac{F_{rgc} A_{ris} h_{ris}}{v_{ris}} \qquad (3.318)$$

3.5.2.1.5 Momentum Balance for Reactor and Main Fractionator Pressure Determination

$$\frac{dP_5}{dt} = 0.833(F_{wg} - F_{V11} - F_{V12} + F_{V13}) \qquad (3.319)$$

$$F_{V12} = k_{12} V_{12} \sqrt{P_5 - P_{atm}} \qquad (3.320)$$

A constant pressure drop, ΔP_{frac}, between reactor and main fractionator is considered; according to this, the reactor pressure is computed by the equation:

$$P_4 = P_5 + \Delta P_{frac} \qquad (3.321)$$

3.5.2.2 Regenerator Model

The mathematical model for the regenerator presents a higher complexity due to the importance of this system in determining the time constant for the entire FCCU.

The regenerator is considered to be divided in two zones: a dense bed zone, and a zone of entrained catalyst (the disengaging zone) (Fig. 3.102).

The dense bed zone consists of two phases: a bubble phase of gaseous reactants and products moving up the bed in plug flow, and a perfectly mixed dense phase containing gases and solid catalyst [128].

Mass transfer occurs between the two phases, but at regenerator temperatures the reaction rates are controlling, rather than mass transfer between the two phases. Since the dense phase is considered to be perfectly mixed, the temperature is assumed uniform in the bed and the gaseous phase in equilibrium with dense phase. Catalyst is present in the zone above the dense bed due to entrainment. The amount of catalyst decreases with the regenerator height. In the entrained catalyst zone the CO combustion is dominant (the amount of catalyst is diminished), having an important heat contribution. The operating conditions are corresponding to CO partial combustion mode.

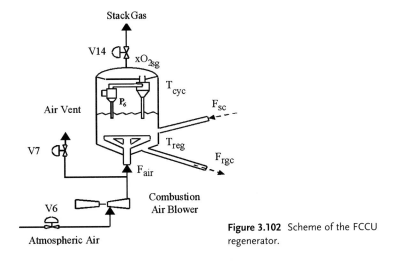

Figure 3.102 Scheme of the FCCU regenerator.

The regenerator model consists of mass and heat balance equations for O_2, CO, CO_2 and coke, and also of heat balance equations for the solid and gaseous phases. These balance equations are correlated with equations describing entrained catalyst (bed characteristics) in the zone above the dense bed, catalyst flow, and pressure in the regenerator.

3.5.2.3 Model of the Catalyst Circulation Lines

For catalyst flow in the spent and regenerated catalyst circulation lines (pipe network), there is assumed to be a steady-state behavior. The dynamics of the lines are considered to be very fast compared to the time constants of other subsystems of the FCCU.

Spent and regenerated catalyst circulation considers a single-phase flow, based on force balance [133]. For the regenerated catalyst line the equation is:

$$
144(P_6 - P_{rb}) + z_{bed} \cdot \varrho_c + (E_{tap} - E_{oil})\varrho_c - \Delta P_{sv,rgc} - \\
- \Delta P_{elb,rgc} - \frac{F_{rgc} \, L_{rgc} \, F_{rgc}}{A_{rgc}^2 \cdot \varrho_c} = 0
$$
(3.322)

and for the spent catalyst line the force balance is given by:

$$
144(P_4 - P_6) + (E_{str} - E_{lift})\varrho_c + \frac{W_r}{A_{str}} - \Delta P_{sv,sc} - \\
- \Delta P_{elb,sc} - \frac{F_{sc} \, L_{sc} \, F_{sc}}{A_{sc}^2 \cdot \varrho_c} = 0
$$
(3.323)

The pressure drop on the slide valves is described by the following equation:

$$
\Delta P_{sv} = \left[\frac{50 \, F_{cat}}{K \cdot A_{sv} \cdot sv} \right]^2 \cdot \frac{144}{\varrho_c}
$$
(3.324)

The pressure drop on other pipe restrictions are given by equations of the type:

$$\Delta P_{elb} = \frac{1}{2} N^* \varrho_c v^2 \qquad\qquad (3.325)$$

Results obtained by dynamic simulation present a good fit with industrial operating data, with simulated variables being situated in a range corresponding to industrial unit behavior (Table 3.15). A comparison between industrial operating data and dynamic simulation results has been carried out for a set of data (one-month period), and has confirmed the main trends of the dynamic behavior both on short and long time scales. Obtaining a better fit is still possible by increasing the complexity of the model, and also often necessary as the properties of the raw material are subject to change.

Table 3.15 Typical operating conditions and values obtained with the simulator.

Process variable	Unit of measurement	Minimum value	Maximum value	Nominal value	Value in the simulator
Air flow rate entering regenerator	$Nm^3\,h^{-1}$	85 000	147 000	98 500	102 514
Air vent flow rate	$Nm^3\,h^{-1}$	0	5500	2500	2510
Regenerator temperature	°C	650	700	682	685.06
Cyclone temperature	°C	677	710	705	708.5
Reactor temperature	°C	490	525	515	516.99
Reactor pressure	Bar	1.2	1.9	1.3	1.279
Regenerator pressure	Bar	1.2	2.8	1.5	1.495
Coke on spent catalyst	Mass fraction	0.009	0.014	0.012	0.01165
Coke on regenerated catalyst	Mass fraction	0.002	0.0045	0.0035	0.00393
CO_2 concentration in flue gas	Volume fraction	0.08	0.16	0.13	0.141
O_2 concentration in flue gas	Volume fraction	0.001	0.008	0.0035	0.00288
CO concentration in flue gas	Volume fraction	0.03	0.08	0.05	0.042
Catalyst inventory in the reactor	tons	30	60	50	55.7

Dynamic simulations reveal the multivariable and nonlinear behavior of the process presenting strong interactions. An inverse response has been noted, denoting multiple paths with opposing effect transmission. Single-loop decentralized control must face strong impediments for such challenging interacting behavior.

The newly developed dynamic simulator offers the possibility of studying different operating regimes induced both by design changes and by changing operation strategies. It also proves to be a valuable tool for investigating the way that different control strategies may be implemented, and their results predicted. Advanced control systems, such as model predictive control algorithms, are based on mathematical models and rely on the dynamic simulator.

3.5.3
Model Predictive Control Results

3.5.3.1 Control Scheme Selection
Based on literature surveys and an analysis of the current industrial FCCU operation, a set of process variables has been selected and considered to have first-role importance in the efficient and safe operation of the unit [128,129,133–137].

The *controlled variables* have been selected to provide, through control, a safe and economic operation [119]. Control of reactor catalyst inventory (reactor level) W_r, provides stabilization of catalyst circulation. It also sets up a buffer for diminishing upsets in coke concentration deposited on the catalyst and for temperature change progressing from the reactor toward the regenerator. The regenerator temperature, T_{reg}, must be maintained at a certain value in order to allow a stable removal of coke from the catalyst. Exceeding the high temperature limit produces a permanent catalyst deactivation; a reduction under a lower limit, leads to coke accumulation on the regenerated catalyst. The reactor temperature, T_r, must be maintained at a certain level to provide a desired maximum conversion of the feed oil. The stack gas oxygen concentration, xO_{2sg}, must be controlled in order to provide a desired coke combustion, preventing both a thermal increase and an inefficient load of the combustion air blower. Maintaining the cyclone temperature, T_{cyc}, under a maximum limit, provides safe thermal operation for the regenerator and for the downstream units (piping and CO boiler).

The *manipulated variables* have been chosen from the set of independent variables possible to be changed from a practical point of view. The main manipulated variables are the spent and regenerated catalyst flow rates that may be changed by regenerated *svrgc* and spent *svsc* slide valve positions. The preheating furnace fuel flow, F_5, is an important manipulated variable with effective action on the thermal balance of the entire unit. The stack gas flow rate from the regenerator, which is changed by stack gas valve position V_{14}, and the air vent flow rate, which is changed by air vent valve position V_7, are other two manipulated variables. The wet gas suction flow rate, which is changed by suction valve position V_{11}, is another manipulated variable considered in the control schemes.

The selected *disturbances* reflect the main upsets that might affect normal operation of the unit: these include main fractionator pressure upset, feed oil coking characteristics (coking rate) upset, and ambient temperature upset [118,133]. The main fractionator pressure disturbance has been included in the simulation by the term ΔP_{frac}, representing the reactor-main fractionator pressure drop. This disturbance reveals the effect of upsets in main fractionator operation, acting on reactor-regenerator system. These main fractionator pressure upsets may appear when: vapor flow is changed as a result of suction flow rate change of wet gas compressor; internal liquid-vapor traffic of the main fractionator is changed due to reboiler and condenser load upset; or by pressure changes induced from the downstream gas recovery unit. An increasing step disturbance has been selected (having +37% amplitude increase and applied at time $t = 500$ s). The coking characteristic of the feed oil, coking rate K_C, was included as a disturbance to study the effect of changes in raw material properties. It was noted that this unmeasured disturbance has a strong effect on the heat balance of the entire unit. A positive step change has been selected for this disturbance (having +3.2% amplitude increase and applied at time $t = 500$ s). The ambient temperature change is a continuous disturbance affecting FCCU on a day-time basis. It consists of combustion air flow rate changes, which introduces low-amplitude upsets into the unit. This disturbance was included as a descending ramp, with negative slope (–16 °C/8 h), applied for 1 h between $t = 300$ s and $t = 3900$ s.

The MPC of the FCCU was designed in a two-level control structure, acting at the top level of the hierarchic control system by cascading the low-level regulatory control loops (usually flow-rate control loops).

A controllability study, based on Relative Gain Array (RGA), has been performed in order to select both the most efficient manipulated variables for changing the controlled variables, as well as determining the best MPC control scheme, among a set of schemes of the same dimensions. The RGA is a measure of interaction between controlled variables, each of the RGA elements denoting the ratio between open-loop and closed-loop gain in decentralized control. This controllability indicator, as a first filter for selecting the best control scheme, proved to be useful not only for decentralized control but also for the multivariable approach [129,138].

Based on this approach, a set of control schemes has been investigated [125,138]. These have different numbers of controlled/manipulated variables: 3×3, 4×4, 5×5, 5×6 schemes, and are presented briefly in Table 3.16.

Table 3.16 Tested control schemes.

Control scheme (Name/dimension)	Controlled variables	Manipulated variables	MPC tuning parameters U_{wt} and Y_{wt}
S1:3 × 3	$W_r\ T_{reg}\ T_r$	svrgc svsc F_5	$U_{wt} = [120\ 120\ 0.8]$ $Y_{wt} = [0.1\ 0.2\ 1]$
S2:3 × 3	$W_r\ T_{reg}\ T_r$	svrgc svsc V_{14}	$U_{wt} = [120\ 120\ 480]$ $Y_{wt} = [0.1\ 0.2\ 1]$
S3:3 × 3	$W_r\ T_{reg}\ T_r$	svrgc svsc V_7	$U_{wt} = [120\ 120\ 600]$ $Y_{wt} = [0.1\ 0.2\ 1]$
S4:3 × 3	$W_r\ T_{reg}\ T_r$	svrgc svsc V_7	$U_{wt} = [75\ 75\ 300]$ $Y_{wt} = [0.1\ 0.2\ 1]$
S5:4 × 4	$W_r\ T_{reg}\ T_r\ xO_{2sg}$	svrgc svsc $V_{14}\ V_7$	$U_{wt} = [30\ 30\ 120\ 120]$ $Y_{wt} = [0.1\ 0.2\ 1\ 0.5]$
S6:4 × 4	$W_r\ T_{reg}\ T_r\ xO_{2sg}$	svrgc svsc $F_5\ V_7$	$U_{wt} = [150\ 150\ 1\ 600]$ $Y_{wt} = [0.1\ 0.2\ 1\ 0.5]$
S7:4 × 4	$W_r\ T_{reg}\ T_r\ xO_{2sg}$	svrg , svsc $V_{11}\ V_7$	$U_{wt} = [150\ 150\ 300\ 600]$ $Y_{wt} = [0.1\ 0.2\ 1\ 0.5]$
S8:5 × 5	$W_r\ T_{reg}\ T_r\ xO_{2sg}\ T_{cyc}$	svrgc svsc $F_5\ V_7\ V_{11}$	$U_{wt} = [150\ 150\ 1\ 600\ 300]$ $Y_{wt} = [0.1\ 0.2\ 1\ 0.5\ 0.5]$
S9:5 × 5	$W_r\ T_{reg}\ T_r\ xO_{2sg}\ T_{cyc}$	svrgc svsc $F_5\ V_7\ V_{11}$	$U_{wt} = [30\ 30\ 0.2\ 120\ 60]$ $Y_{wt} = [0.1\ 0.2\ 1\ 0.5\ 0.5]$
S10:5 × 5	$W_r\ T_{reg}\ T_r\ xO_{2sg}\ T_{cyc}$	svrgc svsc $F_5\ V_7\ V_{14}$	$U_{wt} = [150\ 150\ 1\ 600\ 600]$ $Y_{wt} = [0.1\ 0.2\ 1\ 0.5\ 0.5]$
S11:5 × 5	$W_r\ T_{reg}\ T_r\ xO_{2sg}\ T_{cyc}$	svrgc svsc $V_{11}\ V_7\ V_{14}$	$U_{wt} = [150\ 150\ 300\ 600\ 600]$ $Y_{wt} = [0.1\ 0.2\ 1\ 0.5\ 0.5]$
S12:5 × 6	$W_r\ T_{reg}\ T_r\ xO_{2sg}\ T_{cyc}$	svrgc svsc $F_5\ V_7\ V_{11}\ V_{14}$	$U_{wt} = [150\ 150\ 1\ 600\ 300\ 600]$ $Y_{wt} = [0.1\ 0.2\ 1\ 0.5\ 0.5]$

3.5.3.2 Different MPC Control Schemes Results

The set of MPC schemes presented in Table 3.16 has been tested in the presence of the three typical described disturbances. Different values have been investigated for the error diagonal weighting matrix Y_{wt}, and also for the manipulated-variable move diagonal weighting matrix U_{wt}, from the MPC quadratic optimization objective [139,140].

Following the results obtained by dynamic simulation, the most favorable MPC control schemes, from each category, are: S1:3 × 3, S5:4 × 4, and S10:5 × 5. From this large set of control schemes of MPC dynamic simulations of FCCU, the representative S5:4 × 4 control scheme results are presented in Figures 3.103 and 3.104 [125,128].

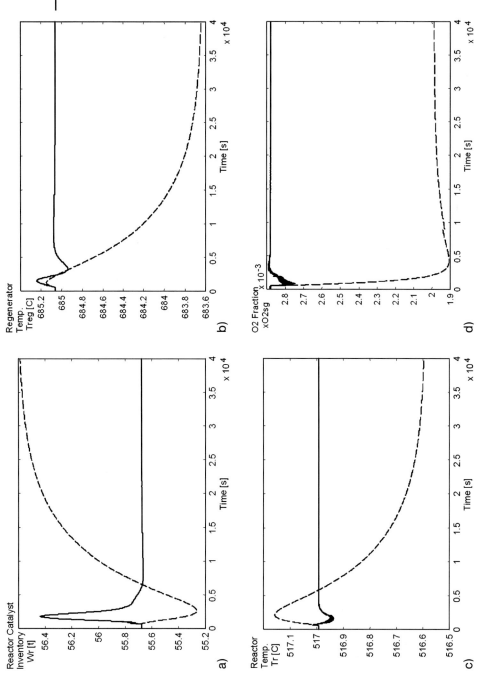

Figure 3.103 MPC simulation results (solid line) in the presence of
K_C disturbance (step increase of coking rate), for S5:4 × 4 control
scheme; disturbed process without control (dashed line). (a) Reactor catalyst inventory; (b) Regenerator temperature; (c) Reactor
temperature; (d) Oxygen fraction.

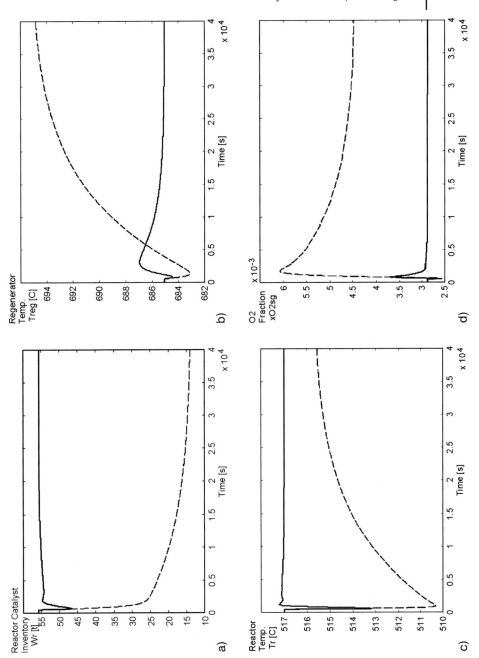

Figure 3.104 MPC simulation results (solid line) in the presence of ΔP_{frac} disturbance (step increase of reactor-main fractionator pressure drop), for S5:4 × 4 control scheme; disturbed process without control (dashed line).
(a) Reactor catalyst inventory; (b) Regenerator temperature;
(c) Reactor temperature; (d) Oxygen fraction.

As can be noted, the S5:4 × 4 control scheme succeeds in counteracting the disturbance effects, and presents small overshoot and short settling times. This control scheme also demonstrates good setpoint following capacity.

The superior behavior of the S5:4 × 4 control scheme, as predicted by the controllability analysis based on RGA values presented in Table 3.17, was confirmed by the dynamic simulation results.

Table 3.17 RGA for S5:4 × 4 control scheme.

	svrgc	*svsc*	V_{14}	V_7
W_r	0.3634	1.3981	−0.8004	0.0390
T_{reg}	2.0118	−0.6095	−0.2298	−0.1725
T_r	0.3946	0.3969	−0.9546	1.1631
xO_{2sg}	−1.7698	−0.1855	2.9848	−0.0296

Compared to S5:4 × 4, the S6:4 × 4 control scheme has inferior control performance, showing higher overshoot and longer response time (especially for the case of K_C disturbance). The S7:4 × 4 control scheme presented unsatisfactory control performance (offset) for all controlled variables in the case of K_C disturbance. In the case of the other investigated disturbances, the control performances of S6:4 × 4 and S7:4 × 4 control schemes were not essentially affected.

Compared to the S1:3 × 3 control scheme, the S5:4 × 4 scheme causes an unimportant increase in overshoot (for the case of K_C performance), but a small decrease in response time was noted. The ability to maintain the stack gas oxygen concentration at a predefined value allows a more efficient FCCU operation due to better use of air blower capacity and to safer operation by the control of the "afterburning" phenomenon. Having an additional variable, compared to the 3 × 3 control schemes, it may be concluded that the S5:4 × 4 scheme is preferable.

Compared to the lower dimension schemes presented before, the 5 × 5 control schemes are characterized by the existence of a higher overshoot and a longer response time, possibly coupled with small offset, though the control performances are not considerably affected.

The S12:5 × 6 MPC scheme did not reveal any improvements compared to the S10:5 × 5 scheme. The advantage of using a control scheme with a higher number of manipulated than controlled variables will become operative when constraints on manipulated variables are imposed. The number of manipulated variable surplus may serve as a supply for the case of operating conditions when one or more of the manipulated variables become restricted.

3.5.3.3 MPC Using a Model Scheduling Approach

The model used to compute manipulated variables is linear, and is obtained by the linearization of the nonlinear model around the operating point [141,142]. The results presented in the previous paragraph were obtained using such a unique model. In order to eliminate errors caused by nonlinearities the authors proposed and investigated the behavior of a control scheme using scheduled linearization. The FCCU linearized model is updated periodically at time moment multiples of 3000 s, starting from $t = 1500$ s. The changing model case has an overall better control performance, particularly for W_r controlled variable (which is affected by the lowest value in the error-weighting matrix) (Fig. 3.105).

The scheduled linearization using a higher frequency of model update did not reveal any significant improvement, for the cases of MPC control in the presence of the investigated disturbances. This may be determined by keeping the operating point relatively close to the setpoint values. As disturbance effects are more important, the updating of the linearized model – at higher and possibly variable frequency – may become necessary.

3.5.3.4 Constrained MPC

Among the most attractive MPC characteristics is the possibility of considering constraints in a direct way. This attribute offers, while specifying FCCU operating and material constraints, the best (in an optimal sense) solution for the control problem. For the SISO case, the requiring of and conforming to constraints is frequently not very difficult. However, for the MIMO case, where interactions are present, the aim of obtaining a desired control performance is usually difficult. On this basis, it is possible that the interest and success of the predictive control algorithm model has been gained in a large number of reported industrial applications [143]. The case of MPC with constraints on manipulated variables has also been investigated [125,128,144].

In order to test this ability, the following potential FCCU malfunction event was simulated. One of the slide valves, the spent catalyst slide valve *svsc*, presents a malfunction when it is unopenable above the upper limit specified by a $svrgc_{sup}$ value of 0.4, and unclosable under the lower limit specified by a $svrgc_{inf}$ value of 0.3. The position of the slide valve during nominal operation was given by $svrgc = 0.35$. This accidental situation raises special problems for the operating personnel in an industrial unit having a traditional (classical) control system. In the case of the model predictive control system, it is sufficient to specify this constraint and to keep the feedback control loops closed until the normal operation regime is restored.

The simulation of MPC behavior for this special operating condition is presented in Figure 3.106, but only for reactor catalyst inventory W_r, controlled variable; other variables exhibit similar behavior with the unconstrained case. The coking rate disturbance K_C has been applied, and the MPC with adaptive model has been simulated. The investigated control scheme is S12:5 × 6. Vector of constraint limits imposed to the manipulated variables is given by $ulim = [0\ 0.3\ 0\ 0\ 0\ 0\ 1\ 0.4\ 1.98\ 0.5$

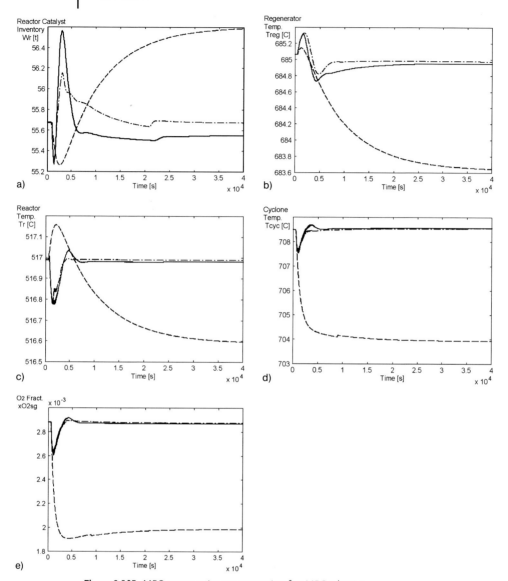

Figure 3.105 MPC comparative representation for: MPC adaptive
model case (dashed-dotted), MPC with unique linearized model
case (solid) and case of disturbed process without control (dashed);
S10:5 × 5 control scheme in the presence of K_C disturbance.
(a) Reactor catalyst inventory; (b) Regenerator temperature;
(c) Reactor temperature; (d) Cyclone temperature; (e) Oxygen fraction.

1 0.8]. The first six values fix the minimum limits, and the last six the maximum
limits allowed for the manipulated variables (in the order they are specified in
Table 3.16).

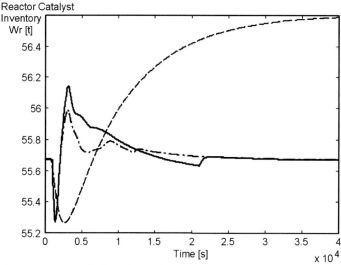

Figure 3.106 Controlled variable W_r for constrained MPC
($svsc_{inf} = 0.3$, $svsc_{sup} = 0.4$), scheme S12:5 × 6 in the presence of
K_C disturbance; unconstrained MPC (solid line), constrained
MPC (dashed-dotted line), disturbed process without control
(dashed line).

As may be observed in Figure 3.106, the MPC control performance is not substantially affected by the occurring constraint. Two of the manipulated variables ($svsc$ and V_7) reached the lower limit values. These limitations do not seem to have any negative impact on the controlled variables, due to the fact that the optimal strategy succeeds in changing the other manipulated variables such that it provides good control performance.

The possibility might also be observed of involving a greater number of manipulated variables than controlled variables, and the potential use of this "excess" of command for cases when constraints on manipulated variables are present.

Based on the results of the present study, it may be considered that this method of MPC application both reveals and sustains the incentives of the MPC algorithm from the perspective of its industrial implementation.

3.5.4
Conclusions

In this section, a new model and dynamic simulator for the FCCU aggregate systems: reactor, regenerator, catalyst circulation lines, preheating system, air blower and wet gas compressor, has been presented. The nonlinear, dynamic and multivariable model has been fitted and subsequently verified with a set of representative operating data originating from an industrial FCCU, showing its complex behavior in response to typical disturbances. It was observed that the disturbance most difficult to reject proved to be the coking rate factor, K_C, although the disturbance considered with the greatest amplitude change was the reactor-main fractionator pressure drop, ΔP_{frac}.

Investigations have been performed by simulation to reveal incentives and limitations for implementing MPC. The most favorable MPC control schemes, for each investigated category, were: S1:3 × 3, S5:4 × 4, and S10:5 × 5. The latter was the most profitable, due to the large number of controlled variables. It was of interest to note that the S12:5 × 6 control scheme (which contained an extra-manipulated variable), in its unconstrained form, does not bring additional quality to MPC. However, when constraints on manipulated variables were present, this approach provided real improvements due to a "surplus" of command that was able to compensate for those manipulated variables limited by constraints. Compared with the traditional decentralized PID control, MPC presents a better control performance based on its multivariable feature, inherent prediction ability, and a capacity to directly handle constraints using even larger numbers of manipulated than controlled variables. A nonlinear MPC method has been proposed and investigated to account for process nonlinearity based on periodic updating of the linearized model used to control action computation. This nonlinear MPC implementation may lead to potential improvements by the use of the dynamic sensitivity analysis in order to obtain adequate MPC tuning.

In practice, MPC implementation is intended to be carried out in a two-layer structure, namely the layer of decentralized PID loops stabilizing the main process variables, and the MPC layer adjusting the setpoints of the underlying regulatory loops.

The benefits of better control performance in FCCU operation consist generally of maintaining the controlled variables very close to the constrained limits, where the optimum operating conditions usually lie.

3.5.5
Nomenclature

A_{rgc}	cross-sectional area of regenerated catalyst pipe (ft²)
A_{ris}	cross-sectional area of reactor riser (ft²)
A_{sc}	cross-sectional area of spent catalyst pipe (ft²)
A_{str}	cross-sectional area of reactor stripper (ft²)
A_{sv}	cross-sectional area of regenerated/spent catalyst slide valve at completely open position (in²)
[COR]	catalyst/oil ratio
C_{cat}	mass fraction of coke produces in the riser
C_{pc}	heat capacity of catalyst (Btu lb⁻¹°F; J kg⁻¹ °K)
C_{pd}	heat capacity of steam (Btu lb⁻¹°F; J kg⁻¹ °K)
C_{pf}	heat capacity of the feed (Btu lb⁻¹°F; J kg⁻¹ °K)
Cp_{fv}	heat capacity of feed vapor (Btu lb⁻¹°F; J kg⁻¹ °K)
C_{rgc} (crgc)	coke fraction on regenerated catalyst (lb coke lb catalyst⁻¹; kg coke kg catalyst⁻¹)

C_{sc} (*csc*) coke fraction on spent catalyst in the stripper (lb coke lb catalyst^{-1}; kg coke kg catalyst^{-1})

C_{sc1} coke fraction on spent catalyst at riser outlet (lb coke lb catalyst^{-1}; kg coke kg catalyst^{-1})

C_1 wet gas production constant (mol lb^{-1} feed; mol kg^{-1} feed)

C_2 wet gas production constant (mol lb^{-1} feed °F; mol kg^{-1} feed °K)

E_{cf} activation energy for coke formation (Btu mol^{-1}; KJ mol^{-1})

E_f activation energy for cracking the feed (Btu mol^{-1}; KJ mol^{-1})

E_g activation energy for cracking gasoline (Btu mol^{-1}; KJ mol^{-1})

E_{lift} elevation of the pipe for spent catalyst, inlet in the regenerator (ft; m)

E_{oil} elevation of feed inlet in the riser (ft; m)

E_{str} elevation of the pipe for spent catalyst outlet from the reactor (ft; m)

E_{tap} elevation of the pipe for regenerated catalyst, outlet from the regenerator (ft; m)

F_{cat} flow rate of spent or regenerated catalyst (t min^{-1})

F_f total feed flow rate (lb s^{-1}; kg s^{-1})

F_{rgc} (*frgc*) regenerated catalyst flow rate (lb s^{-1}; kg s^{-1})

F_{sc} (*fsc*) spent catalyst flow rate (lb s^{-1}; kg s^{-1})

F_{V11} flow through wet gas compressor suction valve V_{11} (mol s^{-1}; mol$_g$ s^{-1})

F_{V12} flow through valve V_{12} (mol s^{-1}; mol$_g$ s^{-1})

F_{V13} flow through valve V_{13} (mol s^{-1}; mol$_g$ s^{-1})

F_{wg} wet gas production in the reactor (mol s^{-1}; mol$_g$ s^{-1})

F_t air flow rate into regenerator (Nm^3 h^{-1})

F_3 fresh feed flowrate (lb s^{-1}; kg s^{-1})

F_4 slurry recycle flowrate (lb s^{-1}; kg s^{-1})

h_{ris} height of the riser (ft; m)

K flow coefficient for the slide valve (0.7)

K_c reaction rate constant for coke production (s^{-1})

k_{r1} reaction rate constant for the total rate of cracking of the feed oil (s^{-1})

k_{r3} reaction rate constant for the rate of cracking gasoline to light gases and coke (s^{-1})

k_{12} wet gas V_{12} valve flow rating (mol s^{-1} psia$^{1/2}$; kg s^{-1} (N m^{-2})$^{1/2}$)

L_{rgc} length of regenerated catalyst pipe (ft; m)

L_{sc} length of spent catalyst pipe (ft; m)

m manipulated variable (input) horizon

m^* factor for the dependence of the initial catalyst activity on C_{rgc}

n model horizon

N exponent for the dependence of C_{cat} on C_{rgc}

N^* integer value representing a constant for pressure drop on catalyst pipes

p prediction horizon

P_{atm} atmospheric pressure (psia; N m^{-2})

P_{rb} pressure at the bottom of the riser (psia; N m^{-2})

P_4 reactor pressure (psia; N m^{-2})

P_5 main fractionator pressure (psia; N m^{-2})

P_6 regenerator pressure (psia; N m^{-2})

R	universal gas constant (ft^3 psia lb^{-1} mol $^{\circ}$R; J mol^{-1} $^{\circ}$K)
sv	spent/regenerated catalyst slide valve position (0–1)
$svsc$	spent catalyst slide valve position (0–1)
$svsc_{inf}$	spent catalyst slide valve lower limit constraint (0.3)
$svsc_{sup}$	spent catalyst slide valve higher limit constraint (0.4)
$svrgc$	regenerated catalyst slide valve position (0–1)
t	time (s)
t_c	catalyst residence time in the riser (s)
T	sampling time (s)
T_{cyc}	regenerator stack gas temperature at cyclone ($^{\circ}$F; $^{\circ}$K)
T_r	temperature of reactor riser outlet ($^{\circ}$F; $^{\circ}$K)
T_{ref}	base temperature for energy balance ($^{\circ}$F; $^{\circ}$K)
T_{reg}	temperature of regenerator bed ($^{\circ}$F; $^{\circ}$K)
T_s	temperature of stripper outlet ($^{\circ}$F; $^{\circ}$K)
T_0	temperature of the feed entering the riser after mixing with the catalyst ($^{\circ}$F; $^{\circ}$K)
T_2	furnace outlet temperature of the feed ($^{\circ}$F; $^{\circ}$K)
$ulim$	vector of constraints imposed to the manipulated variables ([0 0.3 0 0 0 1 0.4 1.98 0.5 1 0.8])
uwt (Γu)	diagonal weighting matrix for the manipulated variable move, in the optimization index
v	catalyst velocity in spent/regenerated pipe (ft s^{-1}; m s^{-1})
v_{ris}	volumetric flowrate in the riser (ft^3 s^{-1}; m^3 s^{-1})
V_{14}	position of the stack gas valve (0–1)
V_7	position of the air vent valve (0–1)
V_{11}	position of the wet gas compressor suction valve (0–1)
V_{12}	position of the flare valve (0–1)
W_r	inventory of catalyst in the reactor (stripper) (lb; kg)
W_{ris}	inventory of catalyst in the riser (lb; kg)
$xO_{2,sg}$	molar ratio of O_2 to air in stack gas (mol O_2/mol air)
$xCO_{,sg}$	molar ratio of CO to air in stack gas (mol CO/mol air)
$xCO_{2,sg}$	molar ratio of CO_2 to air in stack gas (mol CO_2/mol air)
y_f	mass fraction of feed oil
y_g	mass fraction of gasoline
ywt (Γy)	diagonal weighting matrix for the error, in the optimization index
z_a	dimensionless distance along riser
z_{bed}	dense bed height (ft; m)

Greek Symbols

α	catalyst deactivation constant (s^{-1})
ΔH_{evp}	heat of vaporizing the feed oil (Btu lb^{-1}; KJ kg^{-1})
ΔH_f	heat of cracking (Btu lb^{-1}; KJ kg^{-1})
$\Delta P_{elb,sc}$	pressure drop on different elements of spent catalyst pipe (psia; N m^{-2})
$\Delta P_{elb,rg}$	pressure drop on different elements of regenerated catalyst pipe (psia; N m^{-2})
ΔP_{frac}	pressure drop across reactor main fractionator (psi; N m^{-2})

ΔP_{sv}	pressure drop on regenerated/spent catalyst slide valve (psi)
$\Delta P_{sv,\,sc}$	pressure drop on spent catalyst slide valve (psia; N m^{-2})
$\Delta P_{sv,\,rgc}$	pressure drop on regenerated catalyst slide valve (psia; N m^{-2})
ϕ_0	initial catalyst activity at riser inlet
λ	ratio of mass flow rate of dispersion steam to mass flow rate of feed oil
ϱ_c	density of catalyst in the dense phase (lb ft^{-3}; kg m^{-3})
ϱ_{part}	settled density of catalyst (lb ft^{-3}; kg m^{-3})
ϱ_{ris}	average density of material in the riser (lb ft^{-3}; kg m^{-3})
ϱ_v	vapor density at riser conditions (lb ft^{-3}; kg m^{-3})
θ	dimensionless temperature in the riser

3.6
Model Predictive Control of the Drying Process of Electric Insulators

3.6.1
Introduction

The production of high-voltage electric insulators incorporates a two-stage batch-drying process. During the first step, the moisture content of the drying product is reduced from 18–20% to 0.4% in special gas-heated chambers. The second step is carried out in high-temperature ovens, in order to achieve a lower moisture content.

The formed clay insulators are placed on a special support and transport frames, and then introduced into the drying chamber. An electric motor-driven multiple fan provides the air flow through the chamber. The air inlet flow rate can be controlled by means of a butterfly valve. Gas and air flow rates are controlled according to a special program, over a period of about 100 h, in order to obtain the desired moisture content and to avoid the risk of unsafe stress in the drying products [145,146]. First, an analytical dynamic model of the process is derived for model predictive control (MPC) purposes, and second, a neural networks model is trained for the same purpose [147–149].[2)3)]

3.6.2
Model Description

Mass and energy balance equations are used to describe the dynamic behavior of the system [150,151]. The main studied outputs of the model are: moisture content of the drying product X, outlet air temperature T_0, and air humidity x_0; the input

2) Parts of the material presented in this section are reprinted from Cristea, M. V., Baldea, M., and Agachi, S. P., Model Predictive Control of an industrial dryer, *Computer-Aided Chemical Engineering*, **8**, 271–276. Copyright 2000, with permission from Elsevier.

3) Parts of the material presented in this section are reprinted from Cristea, M. V., Roman, R., and Agachi, S. P., Neural networks based MPC of the drying process. *Computer-Aided Chemical Engineering*, **14**, 389–394. Copyright 2003, with permission from Elsevier.

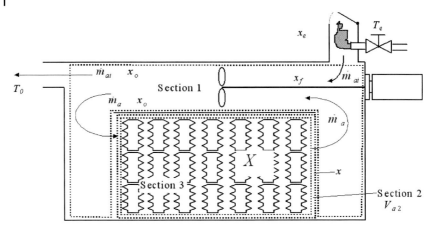

Figure 3.107 Schematic diagram of the drying chamber.

variables studied are: natural gas flow rate $\dot{V}_F$ and mass flow rate of fresh air $\dot{m}_{ai}$. The chamber is divided into three sections, as shown in Figure 3.107. Section 1 represents the air volume within the drying chamber, and section 2 the direct surroundings of the drying product. Section 3 represents the drying product itself.

The mass balance of steam within section 1 and section 2 is described by the equations:

$$\dot{m}_{ai} \cdot x_f + \dot{m}_a \cdot x - (\dot{m}_a + \dot{m}_{ai}) \cdot x_o = V_{a\,Ch} \cdot \varrho_a \cdot \frac{dx_o}{dt} \tag{3.326}$$

$$\dot{m}_a \cdot (x_o - x) - m_S \cdot \frac{dX}{dt} = \frac{d}{dt}(V_{a2} \cdot \varrho_a \cdot x) \tag{3.327}$$

The last term of the equation can be neglected, which results in the differential equation:

$$\frac{dX}{dt} = (x_o - x) \cdot \frac{\dot{m}_a}{m_S} \tag{3.328}$$

By differentiating Eq. (3.328) and assuming that $d^2X/dt^2 \approx 0$, Eq. (3.326) becomes:

$$\frac{dx}{dt} = \frac{1}{V_{a\,Ch} \cdot \varrho_a} \cdot \left(\dot{m}_{ai} \cdot x_f + \dot{m}_a \cdot x - (\dot{m}_a + \dot{m}_{ai}) \cdot x_o\right) \tag{3.329}$$

In Section 3, the behavior of the drying good is described with a normalized diagram by [152,153]:

$$\frac{dX}{dt} = -\frac{\dot{m}_{st}}{m_S} \cdot A_S \tag{3.330}$$

The drying velocity for experimentally determined diagrams characterizes the three periods of the drying process for hygroscopic material, and is normalized according to the following equations [153]:

$$\dot{v}(\eta) = \frac{\dot{m}_{st}}{\dot{m}_{stI}} \qquad\qquad \eta = \frac{X - X_{equ}}{X_c - X_{equ}} \qquad\qquad (3.331)$$

It is assumed that X_c is constant, and does not depend on the drying conditions, and that X_{equ} depends only on relative humidity of the air, and not on other factors. It is also assumed that all diagrams of the drying velocity, for different drying conditions, are geometrically similar. The equilibrium humidity X_{equ} in dependence of the relative air humidity φ, for clay, was considered by a correlation equation. The saturation humidity of the air, x_{sat}, is dependent on the temperature T_o. For low partial pressures of steam, $\dot{m}_{stI}$ was considered according to equation:

$$\dot{m}_{stI} = k \cdot (x_{sat} - x) \qquad\qquad (3.332)$$

with the mass transfer coefficient k determined by experimental data.

Two energy balance equations, for the chamber and for the burner section, are used to describe the outlet temperature change:

$$\dot{m}_{ai} \cdot \left(c_{pa} \cdot (T_i - T_o) + x_f \cdot \left(h_v + c_{pst} \cdot T_i \right) - x_o \cdot \left(h_v + c_{pst} \cdot T_o \right) \right) +$$
$$+ m_s \cdot \frac{dX}{dt} \cdot \left(h_v + c_{pst} \cdot T_o \right) - C_A A_{Ch} (T_o - T_e) = \qquad (3.333)$$
$$= V_{a\,Ch} \cdot \varrho_a \cdot \left(\left(c_{pa} + x_o \cdot c_{pst} \right) \cdot \frac{dT_o}{dt} + c_{pst} \cdot \frac{dx_o}{dt} \cdot T_o \right)$$

$$\dot{m}_{ai} \cdot \left(c_{pa} \cdot T_e + x_e \cdot \left(h_V + c_{pst} \cdot T_e \right) \right) +$$
$$+ \left(c_{pF} \cdot T_e + H_F \right) \cdot \frac{M_F \cdot p_F}{R \cdot (T_e + 273)} \cdot \dot{V}_F = \qquad (3.334)$$
$$= \dot{m}_{ai} \cdot \left(c_{pa} \cdot T_i + x_f \cdot \left(h_V + c_{pst} \cdot T_i \right) \right)$$

A dynamic sensitivity analysis was carried out on this model, indicating the most important parameters and manipulated variables [154]. According to this analysis, these are: mass transfer coefficient k; heat transfer coefficient of chamber walls C_A; heating power of natural gas H_F; mass of the drying product m_S (clay without humidity); environment temperature of the inlet air T_e; environment humidity of the inlet air x_e; volume of the drying chamber V_{Ch}; surface of the drying chamber A_{Ch}; surface of the drying product A_S; critical humidity of clay X_C; and the specific heat of natural gas c_{pF} [155]. The scaled dynamic sensitivity analysis of the output variables with respect to the studied inputs indicated that the natural gas flow rate was the most important manipulated variable (about 10-fold more so than the mass flow rate of fresh air) [154]. The control system was designed accordingly.

3.6.3
Model Predictive Control Results

The first MPC approach obeys the current control practice – that is, driving the moisture content from the product by controlling the air temperature inside the chamber. Usually, the desired decreasing profile of the drying product moisture content is obtained by imposing an increasing ramp-constant profile on the air temperature [145,153]. A comparison was made between traditional PID control and MPC control of the air temperature. Both MPC [147,156] and PID (with anti-windup) control algorithms were implemented in a MISO structure with two manipulated variables: gas flow rate, and air flow rate. For PID, an air:natural gas ratio flow rate control was used. First, the setpoint following capacity was tested in the absence of any disturbances, after which performance testing was carried out for three significant disturbances that typically occur in industrial practice: a 10 °C drop in inlet air temperature T_e (from 16 °C to 6 °C); a 10% drop in the heating power capacity of natural gas, H_F; and a 20% rise in inlet air moisture content, x_e. All three disturbances were introduced as steps at time $t = 110\,000$ s. The simulation results (for the case of the heating power disturbance) are presented in Figure 3.108; this shows the response of the controlled variable over the entire time interval, and a detailed representation of the period when the disturbance acts and is eliminated.

The results revealed a very good behavior, particularly for MPC control. Although both methods exhibited good control ability, the setpoint tracking performance showed a zero-offset behavior for the MPC, whereas PID control proved to be less accurate (mainly for the ramp sections of the setpoint function). The control in the presence of the disturbance emphasizes the superior characteristics of the MPC, with shorter response time and less overshoot compared to the PID control.

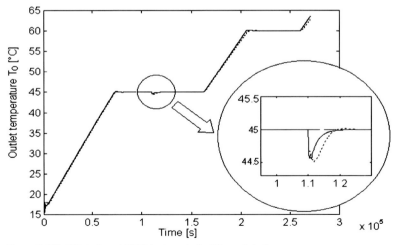

Figure 3.108 PID (····) and MPC (——) control of the outlet air temperature, for a given setpoint

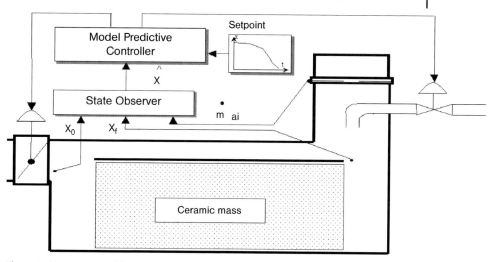

Figure 3.109 Structure of the control system for direct moisture content control.

Taking into account that the target variable – the moisture content of the product – is not measured directly, an inferential state observer is proposed for its estimation. The data provided in this way are used for direct MPC of the moisture content of the drying product (Fig. 3.109).

The selection of a setpoint for moisture content is based on practical and theoretical considerations related to the evolution of product drying rate. The

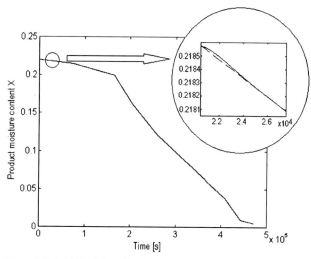

Figure 3.110 MPC of the inferred moisture content of the drying product (——), for a given setpoint (– –) when the heating power disturbance of the fuel, H_F, occurs.

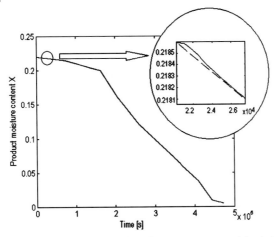

Figure 3.111 MPC of the inferred moisture content of the drying product (—), for a given setpoint (– –) when the inlet air temperature disturbance, T_e, occurs.

conditions stated by the above-mentioned considerations are best fulfilled by a decreasing, seven-segment ramp function, which is actually used as setpoint. Simulations were conducted using this control structure, and the results (disturbances applied at $t = 20\,000$ s) are presented in Figures 3.110 and 3.111.

Again, the setpoint tracking performance was very good. Moreover, the offsets introduced by the disturbances were rapidly eliminated by MPC, and with acceptable deviations from the desired trajectory. The MPC controller was tuned according to the dynamic sensitivity analysis, and based on the maximum allowed variation of both the controlled and manipulated variables.

The applied model predictive algorithm had some special features that made it more effective:

- it had an excess number of manipulated variables over controlled variables;
- in order to achieve the desired control performance, a constrained form of the MPC algorithm was used; and
- the linear model used by the MPC controller is periodically updated to account for the nonlinear behavior of the process [157,158].

3.6.4
Neural Networks-Based MPC

3.6.4.1 Neural Networks Design and Training

Building the artificial neural networks (ANN) model has been the first step in performing the ANN-based Nonlinear MPC (NMPC). The ANN model of the dryer has been developed to serve two goals:

- To provide information on the time evolution of target variables; this is inherently needed for prediction in the NMPC algorithm.

- To infer the moisture content of the drying product, based on available measured variables; this model is used later for the ANN observer-based NMPC.

The ANN-developed model has complementary properties of requiring reduced computation effort, and supplying the algorithm with speed necessary for real-time implementation [149,159].

The structure of the ANN consists of two layers of neurons having as ANN-inputs the natural gas flow rate, the moisture content of the drying product, the outlet air temperature, and the chamber air humidity (each considered with values at the current sampling time, t). The last three variables are the state variables of the process. The ANN-outputs are the same three state variables, but considered at the next sampling time $t+\Delta t$. The trained ANN is designed to predict the behavior of the state variables one step into the future. Applied repeatedly, the dynamic ANN predicts the time evolution of the state variables over a desired time horizon.

Since the drying of electric insulators is performed in batch-wise manner, the process demonstrates in particular a lack of steady states. Hence, according to this behavior, the training procedure of the ANN has been achieved on the basis of a specially prepared set of training data that were chosen in accordance with industrial control practice. The ANN architecture employed was multilayer feed-forward in nature, with the backpropagation training algorithm used to compute the network biases and weights. Two layers of neurons were considered; these had the tan-sigmoid transfer function for the hidden layer, and the purelin transfer function for the output layer. The quasi-Newton Levenberg–Marquardt algorithm was used to train the ANN, and an early stopping method was applied to prevent ANN overfitting and to improve generalization [160].

A good training performance was achieved, as indicated by the correlation coefficients between the training data set (targets) and the ANN simulation data set (ANN response) being close to 1.

After testing, the trained ANN was used to simulate the drying process, having imposed different drying programs compared to those used for training. The favorable fit was also preserved for testing subsets of data showing a good generalization property of the ANN. The prediction capability of the ANN was subsequently exploited for observer-based nonlinear model-based predictive control.

3.6.4.2 ANN-Based MPC Results

The NMPC structure obeys current control practice – that is, driving the moisture from the product by controlling the air temperature inside the chamber. Usually, the desired decreasing profile of moisture content is achieved by imposing an increasing ramp-constant profile on the air temperature. The ANN-based Nonlinear Model Predictive Controller uses the previously trained ANN to perform its prediction tasks. Step response models, simulated by the ANN, were used in either a single-model approach or a multiple-models approach. The latter uses updated models for each of the ramp-constant segments of the drying program.

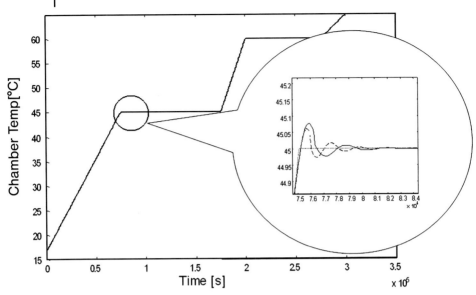

Figure 3.112 Setpoint (dotted line) following ability of the NN based MPC for single (solid line) and multiple (dashed line) model approach.

First, the setpoint following capacity was tested in the absence of any disturbances. The results of a simulation for drying chamber air temperature control are presented in Figure 3.112.

The results revealed a very good behavior, particularly for the NMPC case using multiple models. The control performance showed a reduced overshoot and short settling time, compared to that achieved with the simple PID control structure [148].

Subsequently, control performance testing was repeated in the presence of the three significant disturbances that typically occur in industrial practice: a 10 °C drop in inlet air temperature, T_e (from 16 °C to 6 °C); 10% drop in the heating power capacity of natural gas, H_F; and a 20% rise in inlet air moisture content, x_e.

All three disturbances were introduced as steps at time $t = 120\ 000$ s. The results of the ANN-based NMPC control, for the second (and most important) disturbance, are shown in Figure 3.113.

The disturbance rejection aptitude of the ANN-based NMPC presents favorable control performance results in terms of the setting time, overshoot and to the zero offset, for all tested disturbances.

Taking into account that the target variable – the moisture content of the product – is not available for direct measurement, a ANN-based state observer is proposed for its estimation. The data provided by the ANN state observer is used for feedback nonlinear model predictive control of the product moisture content.

The time-dependent setpoint selection for the moisture content is based on practical and theoretical considerations related to the time evolution of the product drying rate. The conditions stated by the above-mentioned considerations are best

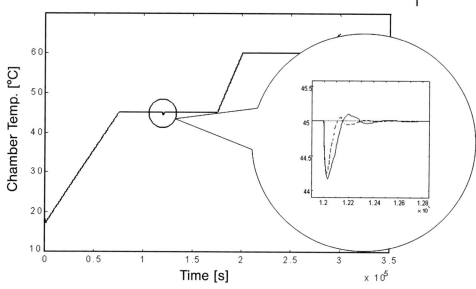

Figure 3.113 Disturbance rejection ability of the NN-based MPC
for the heating power capacity H_F drop, single (solid line) and
multiple (dashed line) model approach.

fulfilled by a seven-segment ramp function, which is actually used as setpoint.
Simulations were conducted using this control structure, and the results for two
10% heating power disturbances, applied at both $t = 3000$ s and $t = 1\ 250\ 000$ s, are
presented in Figure 3.114.

The simulation results for this control structure show both good setpoint
following and disturbance rejection capability. Although a slight decrease in
control performance quality was noted, this change represented a reduced influence on overshoot, response time, and offset. The ability to directly control the
product moisture content may be highly beneficial.

The applied model predictive algorithm has certain special features that render it
more effective: it operates with constraints on manipulated variables and controlled
variables; in order to obtain the feasible control performance, a nonlinear form of
the MPC algorithm was used; and the NMPC controller was tuned with dynamic
sensitivity analysis and is based on the maximum allowed variation of both the
controlled and manipulated variables.

3.6.5
Conclusions

First, the proposed MPC for batch-drying electric insulators proved to be a good
strategy for controlling the drying process. Its high performance was due to the
direct control of product moisture content based on a first principle state observer,
to updating of the model of the process on which MPC relies, and to the optimal

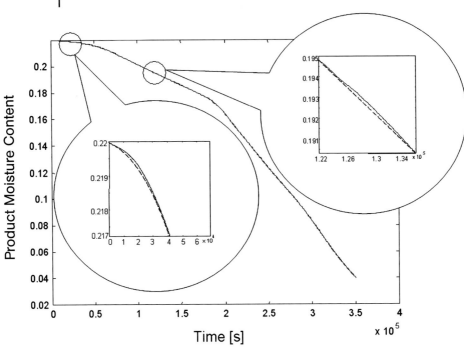

Figure 3.114 Setpoint (dashed line) following and disturbance rejection ability for direct moisture content control using NN-based state observer and NN NMPC (solid line).

manipulation of both the inlet air and gas flow rates. Results obtained simulating MPC control revealed very good setpoint tracking performance as well as effective disturbance rejection.

Second, the proposed ANN-based observer of product moisture content, coupled with the ANN-based model used for NMPC, proved to be a better strategy for controlling the drying of electric insulators. The results obtained when simulating ANN-based NMPC again revealed good setpoint tracking performance, and effective disturbance rejection. This high performance was due to the direct control of product moisture content relying on the ANN-based state observer, to the calculation speed provided by the ANN-based model used in the NMPC algorithm, and to the optimal manipulation of both the inlet air and gas flow rates. The ANN-based model offers the incentives of capturing the intrinsic behavior of the drying process which otherwise was difficult to describe using first principle models.

The use of this method should be conducted with care with regard to the quality of the training set data. An increased confidence in these data may be obtained by expunging the outliers, filtering the data presenting errors (during the ANN simulation steps), and repeating the ANN training procedure, both preceded by careful analysis of the feasibility of the training data. In the industrial plant, this approach may lead to increased energy efficiency, higher productivity, and better product quality.

3.6.6
Nomenclature

A	area of the surface	(m^2)
C_A	heat transfer coefficient of chamber walls	$(W\ (m^2\ °C)^{-1})$
c_p	specific heat	$(J\ (g\ °C)^{-1})$
h_V	heat of vaporization (latent heat)	$(J\ g^{-1})$
H	enthalpy flux	(W)
k	mass transfer coefficient	$(kg\ (s\ m^2)^{-1})$
m	mass (as far as masses of air are concerned, mass of dry air)	(kg)
$\dot{m}$	mass flux	$(kg\ s^{-1})$
$\dot{m}_{st}$	drying rate	$(kg\ (s\ m^2)^{-1})$
p	pressure	(bar)
R	gas constant	$(J\ mol^{-1}\ K)$
T	temperature	$(°C)$
V	volume	(m^3)
X	humidity of drying good – mass of water per mass of dry substance	
		$(kg\ kg^{-1})$
x	humidity of air – mass of water per mass of dry air	$(kg\ kg^{-1})$
η	normalized humidity of drying product	$(\ )$
φ	relative humidity of the air $\varphi = x/x_{sat}$	$(\%)$
ϱ	density	$(kg\ m^{-3})$
$\dot{v}$	normalized drying rate $\dot{v} = \dot{m}_{st}/\dot{m}_{stI}$	$(\ )$

Indices

a	air
c	critical
Ch	chamber
e	environment
equ	equilibrium
I	in
f	fresh
F	fuel
o	out
S	sample
I, II, III	number of drying period
sat	saturation
st	steam

3.7
The MPC of Brine Electrolysis Processes

3.7.1
The Importance of Chlorine and Caustic Soda

Chlorine is essential to the world's chemical industry, with more then 50% of all chemical processing depending on this element. Chlorine was discovered in 1774 by the Swedish chemist Carl Wilhelm Scheele. During the past century, industrial users have identified vast numbers of ways to take advantage of chlorine's useful properties in processes and products. Chlorine is a key building block of modern chemistry by being used in three principal ways: (i) directly (e.g., to disinfect water); (ii) as a raw material for chlorine-containing products (e.g., plastics, pharmaceuticals, pesticides); and (iii) as an intermediate to manufacture nonchlorinated products [161].

Chlorine is produced by passing an electric current through a brine solution (common salt dissolved in water). Essential co-products are caustic soda (sodium hydroxide) and hydrogen.

Caustic soda (sodium hydroxide) is an alkali, and is widely used in many industries, including gold mining, food processing, textile production, soap and other cleaning agent production, water treatment, and effluent control [162].

The annual worldwide production of caustic soda is about 45 million tons. It is used to produce a broad range of inorganic chemicals, and also for general manufacturing, mineral processing, and water treatment [163].

3.7.2
Industrially Applied Methods for Brine Electrolysis

Chlorine has been manufactured industrially for many years. During this time, the industry's firm commitment to the best safety, health and environmental practices has ensured continuous improvement. There are three main processes for the industrial manufacture of chlorine [162–164]:
- mercury cell process;
- diaphragm cell process; and
- ion-exchange membrane (IEM) cell process.

The main advantages and disadvantages of the industrially applied methods for brine electrolysis are presented in Table 3.18.

Table 3.18 Advantages and disadvantages of the three
chlor-alkali processes.

Process	Advantages	Disadvantages
Mercury process	• Simple brine purification; • high-purity chlorine and hydrogen; • 50 wt% caustic direct from the cell.	• Use of toxic mercury; • expensive cell operation; • large floor space; • costly environmental protection.
Diaphragm process	• Use of well brine; • low electrical energy consumption.	• Use of asbestos; • high steam consumption for caustic concentration in expensive multistage evaporators; • low-purity caustic; • low chlorine quality.
Membrane process	• Low energy consumption; • high-purity caustic; • insensitivity to cell load variation and shutdowns; • inexpensive cell operation.	• High-purity brine; • high cost of membrane.

According to the data presented in Table 3.18, it can be concluded that the most advantageous methods of industrial chlorine production are the mercury process and the IEM process. Because these two processes are widely used, we present in the following sections some aspects involved in the MPC of these procedures. In particular, the development of mathematical models for the processes is presented.

3.7.3
Mathematical Model of the Mercury Cell

Mathematical modeling of electrochemical reactors is difficult due to:
• the complexity of phenomena;
• the diversity of reactor types;
• the reduced possibility of following the process in a run; and
• the absence in the literature of data related to the process.

Taking into account these aspects, few references were identified in the literature relating to brine electrolysis modeling [165–168]. The study of the dynamic behavior of mercury cell is important for a better understanding of the process, as well for the development of an efficient control method. For this purpose, a dynamic model of a De Nora cell from Oltchim Ramnicu-Valcea has been developed [169].

A schematic view of a mercury cell is provided in Figure 3.115.

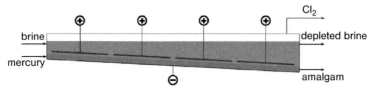

Figure 3.115 Schematic diagram of the mercury cell.

The main equipment of the mercury cell process is the mercury cell itself. This is rectangular in shape, with a width of 1.5–2 m and a length of 15–20 m. The bottom of the cell has a slight slope of 1–3 degrees to the end of the cell in order to allow a smooth flow of mercury on the cell bottom. The anode rods are made from titanium activated with ruthenium oxide (RuO_2) and titanium dioxide (TiO_2). In each cell there are 22 anode lines, each line with three grates of anode rods. The anode lines are grouped into four anode frames that lie parallel to the cell bottom and can be moved up and down by the control system. The depleted brine is recirculated after a resaturation phase.

Chlorine is collected at the top of the mercury cell, while the formed $Na(Hg)_x$ amalgam is guided to the decomposer, where it is washed with warm water in the presence of graphite. Subsequently, sodium hydroxide solution of high purity and hydrogen are obtained due to decomposition of the amalgam. The mercury collected at the bottom of the decomposer is then pumped back to the electrochemical cell and the process is restarted.

The electrochemical process is conducted at high current densities (>10 KA m^{-2}) in order to achieve a high current efficiency for the process.

The main reactions in the mercury cell are as follows [164,166,169]:
- At the anode:
  - Main reaction:
    $2NaCl \rightarrow 2Na^+ + Cl_{2(g)} + 2e^-$
  - Secondary reactions (with impurities from brine):
    $4NaOH + 3Cl_2 \rightarrow 3NaCl + NaClO_3 + 2HCl + H_2O$
    $Na_2CO_3 + 2Cl_2 + H_2O \rightarrow 2HOCl + 2NaCl + CO_{2(g)}$
    $NaHCO_3 + Cl_2 \rightarrow HOCl + NaCl + CO_{2(g)}$
    $Na_2SO_3 + Cl_2 + H_2O \rightarrow Na_2SO_4 + 2HCl$
  - O_2 discharge:
    $4OH^- \rightarrow O_{2(g)} + 2H_2O + 4e^-$
- At the cathode:
  - Main reaction:
    $2Na^+ + 2e^- \rightarrow 2Na$
    $2Na + 2xHg \rightarrow 2Na(Hg)_x$

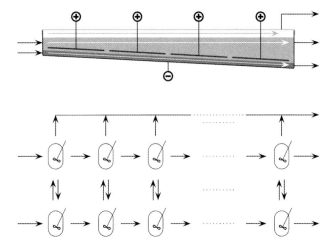

Figure 3.116 Internal structure of the mathematical model for the mercury cell.

3.7.3.1 Model Structure

Taking into account the actual structure of the De Nora mercury cell, the model structure is provided by a series of continuously stirred tank reactors (CSTRs) (Fig. 3.116) [166,169]. Each pair of CSTRs corresponds to an anodic line, representing the discrete unit of the model.

3.7.3.2 The Main Equations of the Mathematical Model

For each anodic line, the mathematical model includes equations for the conservation of mass, energy and impulse, as well as the equation of voltage balance which is specific to the electrochemical processes.

The mass balance equations for the anodic line j are [166,169]:

- Global mass balance:

$$F_{j-1}\varrho_{j-1} - F_j\varrho_j - F_{gaz,j}\varrho_{gas,j} - N_{Na,j}M_{Na} =$$
$$= \frac{\mathrm{d}}{\mathrm{d}t}\left(dz\,l\left(h_{c,j} - h_{am,j}\right)\varrho_j\right) \tag{3.335}$$

- For Cl⁻:

$$F_{j-1}\left(c_{NaCl,j-1} + c_{HCl,j-1}\right) - F_j\left(c_{NaCl,j} + c_{HCl,j}\right) - 2N_{Cl_2,j} =$$
$$= \frac{\mathrm{d}}{\mathrm{d}t}\left(c_{NaCl,j}dz\,l\left(h_{c,j} - h_{am,j}\right)\right) \tag{3.336}$$

- For H⁺:

$$F_{j-1}c_{HCl,j-1} - F_jc_{HCl,j} + N_{HCl,j} = \frac{\mathrm{d}}{\mathrm{d}t}\left(c_{HCl,j}dz\,l\left(h_{c,j} - h_{am,j}\right)\right) \tag{3.337}$$

- For amalgam:

$$F_{am,j-1}\varrho_{am,j-1} - F_{am,j}\varrho_{am,j} + N_{Na,j}M_{Na} = \frac{\mathrm{d}}{\mathrm{d}t}\left(dz\,l\,h_{am,j}\varrho_{am,j}\right) \tag{3.338}$$

- For Na^+ in amalgam:

$$F_{am,j-1}c_{Na,j} - F_{am,j}c_{am,j} + N_{Na} = \frac{\mathrm{d}}{\mathrm{d}t}\left(dz\,l\,h_{am}c_{am,j}\right) \tag{3.339}$$

The energy balance equations for the volume j are [166,169]:
- For the brine:

$$\begin{aligned}
F_{j-1}c_{p,j-1}\varrho_{j-1}T_{j-1} - F_j c_{p,j}\varrho_j T_j + I_j E_{b,j}k - \Delta H_{a,j}(N_{Cl_2} + N_{O_2}) - \\
-2K_{T,l}\left(h_c - h_{am,j}\right)dz\left(T_j - T_{ext}\right) - K_{T,am}l\,dz\,c_{Hg}\left(T_j - T_{am,j}\right) - \\
-Q_{vap}\varrho_{H_2O}r_{H_2O} - \sum_i^{Cl_2,H_2H_2O} n_i M_i c_{p,i} = \frac{\mathrm{d}}{\mathrm{d}t}\left(T_j l\,dz\,\varrho_j c_{p,j}(h_c - h_{am})\right)
\end{aligned} \tag{3.340}$$

- For amalgam:

$$\begin{aligned}
F_{am,j-1}c_{p,am,j-1}\varrho_{am,j-1}T_{am,j-1} - F_{am,j}c_{p,am,j}\varrho_{am,j}T_{am,j} + I_j E_{B,j}(1-k) + \\
+\Delta H_{c,j}N_{Na,j} + K_{T,am}l\,dz\,c_{Hg}\left(T_j - T_{am,j}\right) - \\
-K_{T,f}\left(2h_{am,l} + l\right)dz\left(T_{am,j} - T_{ext}\right) = \frac{\mathrm{d}}{\mathrm{d}t}\left(T_{am,j}l\,dz\,h_{am,j}c_{p,am,j}\right)
\end{aligned} \tag{3.341}$$

These equations are completed with the voltage balance equation [170]:

$$E_B = E + \eta_a - \eta_c + \Delta\varepsilon_\Omega + \Delta\varepsilon_c \tag{3.342}$$

where:

$$E = \varepsilon_a - \varepsilon_c$$

The model includes equations for bubble development [171] and amalgam flow, as well as for the physical properties of the fluids (gas, liquid: brine and amalgam) presented in Table 3.19.

Within this structure, the mathematical model of De Nora cell includes 154 differential equations and more than 440 nonlinear algebraic equations.

The model was tested based on data provided from more than 100 experiments, obtained in the mercury cell plants from *Oltchim* Râmnicu-Vâlcea and *Chimcomplex* Borzeşti (both in Romania).

Table 3.19 Equations of the mathematical model for the mercury cell.

Description	Equation	Reference(s)
• Standard anodic potential for the main anodic reaction	$\varepsilon^0_{Cl^-/Cl_2} = 1.47252 + 4.82271 \cdot 10^{-4} T - 2.90055 \cdot 10^{-6} T^2$	172
• Reversible anodic potential	$\varepsilon_a = \varepsilon_{Cl^-/Cl_2} = \varepsilon^0_{Cl^-/Cl_2} + \dfrac{RT}{2\mathfrak{F}} \ln p_{Cl_2} - \dfrac{RT}{\mathfrak{F}} \ln a_{Cl^-}$	172
• Activity of Cl^-	$a_{Cl^-} = c_{Cl^-} (0.777 + 0.081(c_{Cl^-} - 4) - 0.0014(T - 353))$	164
• Standard cathodic potential for the main cathodic reaction	$\varepsilon^0_{Na^+/NaHg} = -1.71122 - 5.57768 \cdot 10^{-4} T - 8.7953 \cdot 10^{-7} T^2$	164
• Reversible cathodic potential	$\varepsilon_c = \varepsilon_{Na^+/NaHg} = \varepsilon^0_{Na^+/NaHg} + \dfrac{RT}{\mathfrak{F}} \ln a_{Na^+} - \dfrac{RT}{\mathfrak{F}} \ln a_{NaHg}$	164
• Anodic overpotential	$\eta_\alpha = a + b \cdot \log j$ where: $a = 1.3322 \cdot 10^{-5} T$ and $b = 1.3212 \cdot 10^{-4} T$	173,174
• Cathodic overpotential	$\eta_c = 0.25 \cdot \eta_a$	164
• Voltage drop on the electrolyte	$\Delta \varepsilon_\Omega = I \cdot R_{el}$	175
• Electrical resistance of the electrolyte	$R_{el} = \dfrac{A}{d} \left(\lambda_{HCl} + \lambda'_{NaCl} \right)$	176
• Conductance of the brine with bubbles	$\lambda'_{NaCl} = \lambda_{NaCl} (1 - f_{gas})^{3/2}$	177
• Voltage drop on contacts and conductors	$\Delta \varepsilon_c = k_c (0.033 + 1.1666 \cdot 10^{-2} i)$	174
• Amalgam layer thickness	$h_{am} = \sqrt[3]{\dfrac{3 \, G_{am} \, \eta_{am}}{lg \, \varrho^2_{am} \, \alpha}}$	178
• Corrections for amalgam thickness (based on experimental data)	$h^m_{am} = h_{am} (c_1 + c_2 (e^{c_3 c_{Na,j}} + c_4 e^{-jc_5} - 1))$	166
• Electrode gap	$d = d_c - h^m_{am}$	
• Gas fraction in the interpolar gap	$f_{gaz} = 1 - \dfrac{d - G_g}{d}$	176,179,180
• Densities	brine $\varrho_{NaCl} = (a + b \cdot c_{NaCl} + c \cdot c^2_{NaCl}) \cdot 10^3$ amalgam $\varrho_{am} = 13500 - 51.111 \cdot (c_{Na} - 2.9348)$	166,168
• Amalgam viscosity	$\eta_{am} = 1.9 \cdot 10^{-3} e^{x_{Na} [0.975 - 3.54 \cdot 10^{-3}(T_{am} - 273)]}$	166

3.7.4

Mathematical Model of Ion-Exchange Membrane Cell

The mathematical model presented in this section is an original model based on information obtained only from the literature. This model has been developed for a Hoechst-Uhde BM cell, and also used by the IEM cell plant from *Chimcomplex Borzeşti* [166,181–184].

This mathematical model is a dynamic model, developed for process control. The IEM cell is illustrated in Figure 3.117.

In the IEM cell process, the anode and cathode are separated by a cation-exchange membrane [161,164,185,186]. Only sodium ions and a little water migrate through the membrane. As for the mercury process, the brine is dechlorinated and recirculated for resaturation with solid salt. The life of the expensive membranes depends on the purity of the brine. Therefore, after initial purification by precipitation-filtration, the brine is additionally purified by ion-exchange of higher valent cations (Ca^{2+}, Ba^{2+}, and Mg^{2+}) [187,188].

The caustic solution leaves the cell at a concentration of 30–40 wt.%, and must be further concentrated. The chlorine content of the sodium hydroxide solution is as low as that from the mercury process. The chlorine gas contains some oxygen, and must be purified by liquefaction and evaporation. The consumption of electric energy with the membrane cell process is the lowest of the three processes (ca. 25% less than that for the mercury process). The amount of steam needed to concentrate the caustic solution is relatively small. Moreover, there are no special environmental problems, and the cells are easy to operate and relatively insensitive to current density changes, allowing greater use of cheaper, off-peak electric power.

The main reactions are as follows [164,166,181]:

- Anode compartment:
  - Main reaction:

    $$2NaCl \rightarrow 2Na^+ + Cl_{2(g)} + 2e^-$$

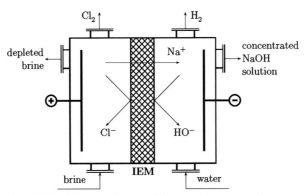

Figure 3.117 Schematic diagram of the ion-exchange membrane (IEM) cell.

- Secondary reactions (with impurities from brine):

$$4NaOH + 3Cl_2 \rightarrow 3NaCl + NaClO_3 + 2HCl + H_2O$$
$$Na_2CO_3 + 2Cl_2 + H_2O \rightarrow 2HOCl + 2NaCl + CO_{2(g)}$$
$$NaHCO_3 + Cl_2 \rightarrow HOCl + NaCl + CO_{2(g)}$$
$$Na_2SO_3 + Cl_2 + H_2O \rightarrow Na_2SO_4 + 2HCl$$

- O_2 discharge:

$$4OH^- \rightarrow O_{2(g)} + 2H_2O + 4e^-$$

- Cathode compartment:
  - Main reaction:

$$2H_2O + 2e^- \rightarrow H_{2(g)} + 2HO^-$$

  - Secondary reaction:

$$Na^+ + HO^- \rightarrow NaOH$$

3.7.4.1 Model Structure

The model which has been considered is formed by a series of CSTRs having the structure shown in Figure 3.118 [166].

The basic element of the model is considered to be the pair of CSTRs situated on the same level in the anode, or respectively cathode, compartment of the cell. The pair of CSTRs includes one ideal reactor for anode, respectively cathode compartment, at the opposite side of the membrane. For a pair of CSTRs, there is a streams structure, as shown in Figures 3.119 and 3.120 [166].

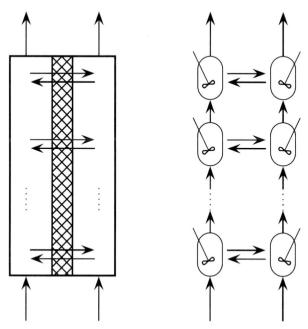

Figure 3.118 Internal structure of the mathematical model.

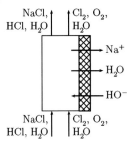

Figure 3.119 Streams structure in the anode compartment.

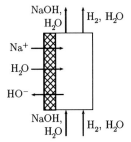

Figure 3.120 Streams structure in the cathode compartment.

3.7.4.2 The Main Equations of the Mathematical Model

For each CSTR, energy, mass, voltage balance equations, equations describing the physical and electrical properties of the phases were considered. For an adequate description of the process, and in order to limit the calculus complexity after multiple simulations, the number of these CSTR pairs was limited to 10. This provided good correspondence with the data acquired from the industrial process.

For each basic element of the model, mass, energy and voltage balance equations were considered as well as equations for physical properties.

The mass balance equations are [166]:

• Global mass balance equation for the anodic compartment:

$$
F_{a,i}\varrho_{a,i} + F_{ga,i}\varrho_{ga,i} + N_{OH^-}M_{OH} - N_{Na}M_{Na} - N_{H_2O}M_{H_2O}-
$$
$$
-F_{a,e}\varrho_{a,e} - F_{ga,e}\varrho_{ga,e} = \frac{d}{dt}\left(V_a\varrho_{a,a}\right)
\tag{3.343}
$$

– For Na$^+$ in the anodic compartment:

$$
F_{a,i}\left(c_{a,i} + c_{NaClO_3,i}\right) - F_{a,e}\left(c_{a,e} + c_{NaClO_3,e}\right) - N_{Na} =
$$
$$
= \frac{d}{dt}\left(V_a\left(c_{a,e} + c_{NaClO_3,e}\right)\right)
\tag{3.344}
$$

– For Cl$^-$ in the anodic compartment:

$$
F_{a,i}c_{a,i} - F_{a,e}c_{a,e} - \frac{I}{\mathbf{F}}r_{Cl_2} + \frac{3}{4}N_{OH} = \frac{d}{dt}\left(V_a c_{a,e}\right)
\tag{3.345}
$$

– For ClO_3^- in the anodic compartment:

$$F_{a,i}c_{ClO_3^-,i} - F_{a,e}c_{ClO_3^-,e} + \frac{1}{4}N_{OH} = \frac{d}{dt}\left(V_a(1 - f_{ga})c_{ClO_3^-,e}\right) \tag{3.346}$$

– For HCl in the anodic compartment:

$$F_{a,i}c_{HCl,i} - F_{a,e}c_{HCl,e} + \frac{1}{2}N_{OH} = \frac{d}{dt}\left(V_a(1 - f_{ga})c_{HCl,e}\right) \tag{3.347}$$

• Global mass balance equation for the cathodic compartment:

$$F_{c,i}\varrho_{c,i} + N_{Na}M_{Na} + N_{H_2O}M_{H_2O} + F_{gc,i}\varrho_{gc,i} - F_{c,e}\varrho_{c,e} - $$
$$-N_{OH}M_{OH} - F_{gc,e}\varrho_{gc,e} = \frac{d}{dt}\left(V_c\varrho_{am,c}\right) \tag{3.348}$$

– For Na^+ in the cathodic compartment:

$$F_{c,i}c_{NaOH,i} + N_{Na} - F_{c,e}c_{NaOH,e} = \frac{d}{dt}\left(V_c c_{NaOH,e}\right) \tag{3.349}$$

The energy balance equations are [166]:
• For the anodic compartment:

$$F_{a,i}\varrho_{a,i}c_{p_{a,i}}T_{a,i} + F_{ga,i}\varrho_{ga,i}c_{p_{ga,i}}T_{a,i} + \alpha_T IE_b - \Delta H_a(N_{Cl_2} + N_{O_2}) - $$
$$-2AK_{T_{ac}}\left(T_{a,e} - T_{c,e}\right) - 2BK_{T_{ext}}\left(T_{a,e} - T_{ext}\right) - F_{a,e}\varrho_{a,e}c_{p_{a,e}}T_{a,e} - $$
$$-M_{H_2O}\left(F_{ga,e}c_{H_2O,e} - F_{ga,i}c_{H_2O,i}\right)r_{H_2O} - F_{ga,e}\varrho_{ga,e}c_{p_{ga,e}}T_{a,e} = $$
$$= \frac{d}{dt}\left(V_a\varrho_{am,a}c_{p_{a,e}}T_{a,e}\right) \tag{3.350}$$

• For the cathodic compartment:

$$F_{c,i}\varrho_{c,i}c_{p_{c,i}}T_{c,i} + F_{gc,i}\varrho_{gc,i}c_{p_{gc,i}}T_{c,i} + (1 - \alpha_T)IE_b - \Delta H_c N_{H_2} - $$
$$-2AK_{T_{ac}}\left(T_{c,e} - T_{a,e}\right) - 2BK_{T_{ext}}\left(T_{c,e} - T_{ext}\right) - $$
$$-F_{c,e}\varrho_{c,e}c_{p_{c,e}}T_{c,e} - M_{H_2O}\left(F_{gc,e}c_{H_2O,e} - F_{gc,i}c_{H_2O,i}\right)r_{H_2O} - $$
$$-F_{gc,e}\varrho_{gc,e}c_{p_{gc,e}}T_{c,e} = \frac{d}{dt}\left(V_c\varrho_{am,c}c_{p_{c,e}}T_{c,e}\right) \tag{3.351}$$

The mathematical model also includes the voltage balance equation [175]:

$$E_B = E + \eta_a - \eta_c + \Delta\varepsilon_\Omega + \Delta\varepsilon_m + \Delta\varepsilon_c \tag{3.352}$$

where:

$$E = \varepsilon_a - \varepsilon_c$$

Other equations of the mathematical model are presented in Table 3.20.

Table 3.20 Equations of the mathematical model for the ion-exchange membrane (IEM) cell.

Description	Equation	Reference(s)
• Anodic standard potential	$\varepsilon^0_{Cl^-/Cl_2} = 1.47252 + 4.82271 \cdot 10^{-4}T - 2.90055 \cdot 10^{-6}T^2$	172
• Reversible potential of the anodic process	$\varepsilon_a = \varepsilon_{Cl^-/Cl_2} = \varepsilon^0_{Cl^-/Cl_2} + \dfrac{RT}{2\mathfrak{F}}\ln p_{Cl_2} - \dfrac{RT}{2\mathfrak{F}}\ln a_{Cl^-}$	172
• Heat of reaction for the anodic process	$\Delta H_a = 147234.94 \cdot 10^3 + 267.37\ T^2$	175
• Reversible potential of the cathodic process	$\varepsilon_c = \varepsilon_{H^+/H_2} = -0.828 - \dfrac{RT}{2\mathfrak{F}}\ln p_{H_2} - \dfrac{RT}{\mathfrak{F}}\ln a_{HO^-}$	175
• Heat of reaction for the cathodic process	$\Delta H_c = 93117.647 \cdot 10^3 - 398.1922\ T^2$	164,175
• Anodic overpotential	$\eta_\alpha = a + b \cdot \log j$ where: $a = 1.3322 \cdot 10^{-5}T$ and $b = 1.3212 \cdot 10^{-4}T$	173
• Cathodic overpotential	$\eta_c = a + b \cdot \log j$ where: $a = 0.02$ and $b = -0.11$	164
• Electrical resistance of the membrane	$\varrho_{mem} = 2.6125 \cdot 10^{-4} - 1.75 \cdot 10^{-6}(T - 273)$	189
• Resistance and voltage drop on the membrane	$R_{mem} = \dfrac{\varrho_{mem}}{A}$ and $\Delta\varepsilon_\Omega = I\ R_{mem}$	190
• Voltage drop on contacts and conductors	$\Delta\varepsilon_c = k_\tau \left(0,033 + 0,0467\ \dfrac{I}{A\ I_{nom}} \right)$	191
• Densities	for the anodic compartment $\varrho_{NaCl} = \left(a + b \cdot c_{NaCl} + c \cdot c^2_{NaCl} \right) \cdot 10^3$ for the cathodic compartment $\varrho_{NaOH} = 1240 - 0.665\left[T - (273 + 60) \right] + \dfrac{c_{NaOH} - 7.44}{0.0366}$	168
• Transport numbers	$t_{Na} = -3,4204 \cdot 10^{-4}c^4_{NaOH} + 1,0531 \cdot 10^{-2}c^3_{NaOH} - 1,1548 \cdot 10^{-1}c^2_{NaOH} + 5,3891 \cdot 10^{-1}c_{NaOH}$ $t_{H_2O} = -1,824 \cdot 10^{-3}c^4_{NaOH} + 6,0262 \cdot 10^{-2}c^3_{NaOH} - 6,8245 \cdot 10^{-1}c^2_{NaOH} + 2,9137\ c_{NaOH}$	189

This structure of the mathematical model includes 70 differential equations and more than 160 nonlinear algebraic equations.

3.7.5
Simulation of Brine Electrolysis

Solving the huge system of differential equations was possible only by using numeric methods. MATLAB/SIMULINK software environment was used for this task. It was possible to determine the inside profile of parameters such as: concentration, temperature, flow, gas fraction, current (current distribution), voltage, pH, electrode coverage, etc., by simulation. A short selection of the simulation results are presented in the following sections.

3.7.5.1   Simulation of the Mercury Cell Process

The simulation results presented in Figure 3.121 show the dynamics of the process in the case of a step change of the feeding brine temperature [166].

The graphics presented an initial inverted response for the majority of the system parameters. For example, the profile for the voltage dynamics we could observed on frames I to IV. This phenomenon could generate severe control problems especially for simpler control systems such as SISO.

3.7.5.2   Simulation of the Ion-Exchange Membrane Cell Process

In the steady state process simulation we obtain the parameters profile along the height of the cell. The graphics in Figure 3.122 present the profile of the sodium hydroxide concentration in the cathode compartment and the profile of the temperature in the anode compartment.

In the case of dynamic simulation, modification of the parameters in time could be calculated in the case of changes in input or state parameters of the cell.

The graphics presented in Figure 3.123 represent the dynamics of the cell voltage, anode compartment temperature, brine concentration and hydroxide concentration at the exit of the cell in the case of feed brine temperature changing from 70 °C to 90 °C.

Also observed in this case was the complexity of the process responses due to complex interactions between the process from the anode and cathode compartments.

3.7.6
Model Predictive Control of Brine Electrolysis

The most comprehensive and powerful model predictive control techniques are based on the optimization of a quadratic objective function, which involves the error between the set point and the predicted outputs. These techniques have been used successfully in commercial process control applications involving MIMO processes. They also can handle inequality constrains on the controlled and manipulated variables.

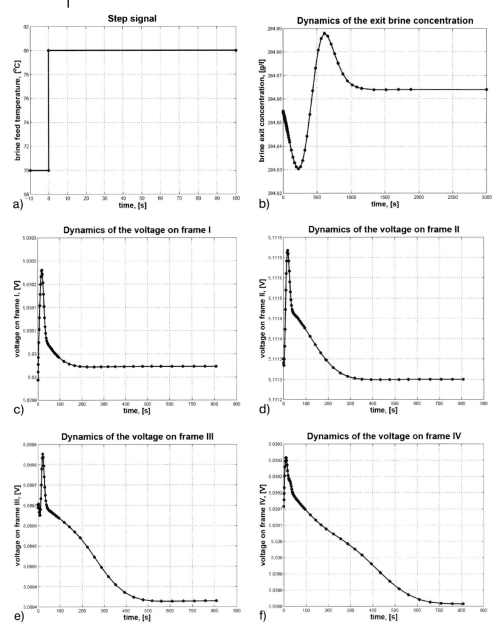

Figure 3.121 (a–f) Simulation results for the mercury cell for a step change in the brine feeding temperature.

NaOH concentration in the cathodic compartment

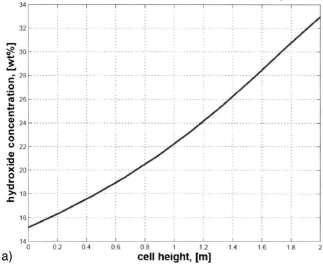

a)

Temperature in the anodic compartment

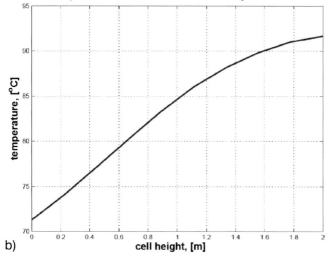

b)

Figure 3.122 Profiles of (a) NaOH concentration and
(b) temperature in the IEM cell obtained by simulation.

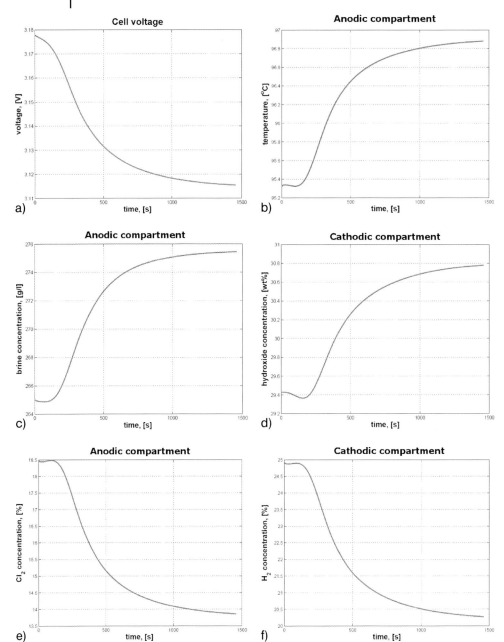

Figure 3.123 (a–f) Dynamics of the IEM cell in the case of feed brine temperature step change from 70 °C to 90 °C at $t = 0$ s.

3.7.6.1 MPC of Mercury Cell

The dynamic responses of the mercury cell (see above) were used to test two different types of control technique: SISO (Single Input/Single Output) control structures using PID controllers; and MIMO (Multiple Inputs/Multiple Outputs) control structures based on Model Predictive Control.

For the SISO control structure as applied to a mercury cell, the following loops were selected:

- Loop I
  - controlled variable: voltage on frame 1
  - manipulated variable: distance between anodic frame 1 and the bottom of the cell
- Loop II
  - controlled variable: voltage on frame 2
  - manipulated variable: distance between anodic frame 2 and the bottom of the cell
- Loop III
  - controlled variable: voltage on frame 3
  - manipulated variable: distance between anodic frame 3 and the bottom of the cell
- Loop IV
  - controlled variable: voltage on frame 4
  - manipulated variable: distance between anodic frame 4 and the bottom of the cell

Four PID controllers were used for the four loops. Controller tuning was made by simulation using the Ziegler–Nichols method. Parameters for these controllers are presented in Table 3.21.

Table 3.21 Controller parameters for mercury cell.

Controller	Type	KR	TI [s]
1	PI	9	16
2	PI	12	16
3	PI	13	16
4	PI	13	16

In the case of MPC of the mercury cell when, using the same controlled variables and manipulated variables, the optimal values for the internal parameters of the controller were determined by simulation, as follows:

- model horizon $T = 14\ 400$ s;
- control horizon $U = 3$;

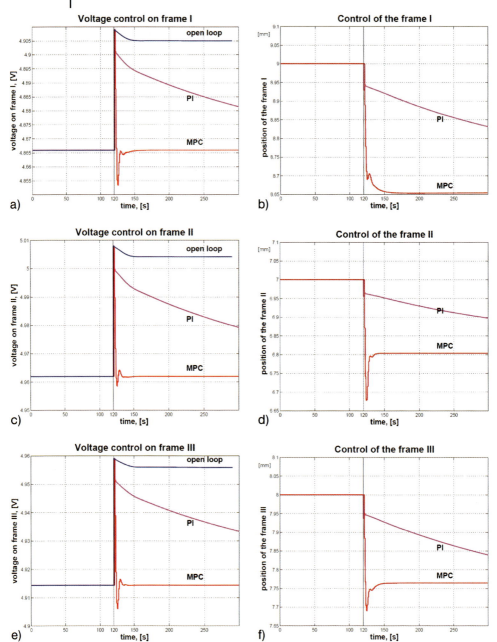

Figure 3.124 (a–h) MPC versus PID for mercury cell. Considered disturbance: modification of cell current load at $t = 120$ s.

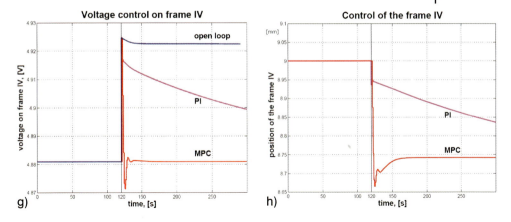

g)

h)

- prediction horizon $V = 8$;
- weighting matrix for predicted errors $W_1 = [300\ 300\ 300\ 300]$;
- weighting matrix for control moves $W_2 = [1\ 1\ 1\ 1]$;
- sampling period $\Delta t = 1$ s.

The controlled variables were subject to the following constraints:

$$y_{min} \leq y \leq y_{max} \tag{3.353}$$

where:

$y_{min} = h_{am} + 0.003$ (mm)

$y_{max} = 4\ y_{min}$

h_{am} = thickness of amalgam layer (mm), computed by the model.

For example, for a correct comparison of the performances of SISO and MIMO control structures, a modification of cell load, from 300 KA to 310 KA, at $t = 120$ s was simulated. The obtained results are presented in Figure 3.124.

It could not be observed from these graphics whether the MPC provided a more effective control than the PI control.

3.7.6.2 MPC of IEM Cell

The dynamic responses of the membrane cell (see above) were used to test two different types of control technique: SISO (Single Input/Single Output) control structures using PID controllers; and MIMO (Multiple Inputs/Multiple Outputs) control structures based on Model Predictive Control.

For the SISO control structure in the case of a membrane cell, the following loops were selected:

- Loop I
  - controlled variable: brine concentration at cell outlet
  - manipulated variable: brine inlet flow
- Loop II
  - controlled variable: caustic soda concentration at cell outlet
  - manipulated variable: caustic soda inlet flow

Two PID controllers were used for these two loops. Controller tuning was achieved by simulation with the Ziegler–Nichols method. Parameters for these controllers are presented in Table 3.22.

Table 3.22 Controller parameters for ion-exchange mercury cell.

Controller	Type	KR	TI [s]
1	PI	15	4500
2	PI	30	5200

In the case of MPC of the membrane cell, when the same controlled variables and manipulated variables were used, the optimal values for the internal parameters of the controller were determined by simulation as follows:
- model horizon $T = 14\,400$ s;
- control horizon $U = 2$;
- prediction horizon $V = 10$;
- weighting matrix for predicted errors $W_1 = [0.05\ 0.05]$;
- weighting matrix for control moves $W_2 = [1\ 1]$;
- sampling period $\Delta t = 1$ s.

The controlled variables are subject to constrain presented by inequalities (3.353). In this case:

$y_{min} = 0$;

$y_{max} = 2\ y_{nom}$

y_{nom} nominal flow for brine/caustic soda, [m^3 s^{-1}].

For a correct comparison of the performances of SISO and MIMO control structures respectively in the case of a membrane cell, a modification of caustic soda concentration at cell inlet, from 10 wt.% to 12 wt.%, at $t = 120$ s, was simulated. The results are presented in Figure 3.125 [192,193].

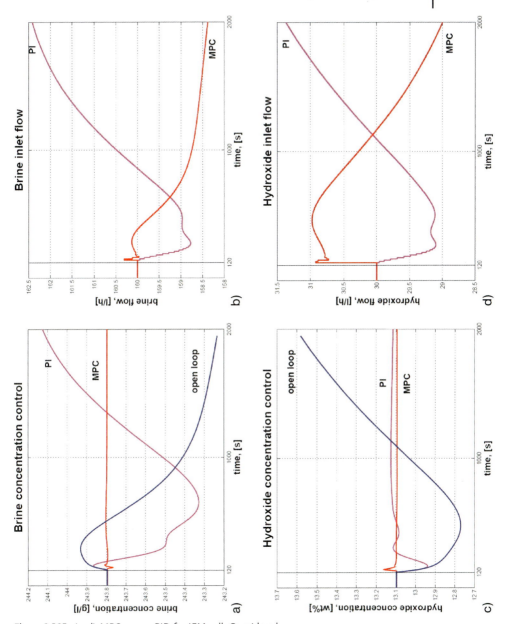

Figure 3.125 (a–d) MPC versus PID for IEM cell. Considered disturbance: modification of caustic soda concentration at cell inlet at $t = 120$ s.

3.7.7
Conclusions

In this section, results related to the modeling and control of industrial brine electrolysis processes (using mercury cell and IEM cell) were presented. The mathematical models for both processes were based on balance equations (mass, energy and voltage balance equations – specific for electrochemical processes), and also included equations presented in the literature or determined by measurements in actual industrial plants.

The nonlinear, dynamic and multivariable models were fitted and then verified with data measured in industrial plants (at Râmnicu-Vâlcea and Borzeşti). Observations made from the simulations indicated that the processes showed a complex behavior in response to typical disturbances.

Control of the processes was studied by considering two different approaches: (i) SISO control using multiple PID controllers; and (ii) by MIMO control based on MPC with 2×2 or 4×4 inputs/outputs. As was observed from the graphics presented, the MPC outperformed the SISO control structures, assuring a better control of the processes.

In conclusion, MPC is a control method that must be taken into consideration also for the control of electrochemical processes.

3.7.8
Nomenclature

Symbols

a	constant
A	cross-sectional area of the membrane
B	side area of the cell
c	molar concentration
c_p	specific heat
dz	length of the volume element
E_B	cell voltage
f_g	gas fraction in the liquid
F	inlet volumic flow
$\mathfrak{F}$	Faraday's number
ΔH	enthalpy
i	current density
I	current intensity
K_T	thermal transfer coefficient
K_c	contacts state coefficient
l	cell width
m	molal concentration
M	molecular mass
N	mole number
p	pressure

r	current efficiency
R	universal gas constant, electric resistivity
t	time, temperature, transport number
T	absolute temperature
V	volume
x, y	molar fraction
z	number of electrons
α	temperature coefficient
α_T	energy repartition coefficient
ε	reversible potential
$\Delta\varepsilon$	voltage drop
γ	activity coefficient
η	overpotential
Λ	conductivity
ϱ	density, resistivity

Subscripts

a	anodic
am	mixture
c	cathodic
$cond$	conductor
e	exit
el	electrolyte
ext	external
g	gas
i	inlet
j	index of the model element
l	liquid
mem	membrane
v	vapor
0	initial, reference

References

1 Nunes, R. W., Martin, J. R., Johnson, J. F., Influence of molecular weight and molecular weight distribution on mechanical properties of polymers, *Polym. Eng. Sci.*, **1982**, *22*, 205–228.

2 Sundaram, B. S., Upreti, S. R., Lohi, A., Optimal control of batch MMA polymerization with specified time, monomer conversion, and average polymer molecular weights, *Macromolecular Theory and Simulations*, **2005**, *14*(6), 374–386.

3 Osakada, K., Fan, L. T., Computation of near-optimal control policies for free-radical polymerization reactors, *J. Appl. Polym. Sci.*, **1970**, *14*, 3065–3082.

4 Ponnuswamy, S. R., Shah, S. L., Computer optimal control of batch polymerization reactors, *Ind. Eng. Chem. Res.*, **1987**, *26*, 2229–2236.

5 Thomas, I. M., Kiparissides, C., Computation of the near-optimal temperature and initiator policies for a batch

polymerization reactor, *Can. J. Chem. Eng.*, **1984**, *62*, 284–291.

6 Takamatsu, T., Shioya, S., Okada, Y., Molecular weight distribution control in a batch polymerization, *Ind. Eng. Chem. Res.*, **1988**, *27*, 93–99.

7 Pana, Y., Jay, H. L., Recursive data-based prediction and control of product quality for a PMMA batch process, *Chemical Engineering Science*, **2003**, *58*, 3215–3221.

8 Park, M.-J., Rhee, H.-K., Property evaluation and control in a semibatch MMA/copolymerization reactor, *Chemical Engineering Science*, **2003**, *58*, 603–611.

9 Boutayeb, M., Darouach, M., Recursive identification method for MISO Wiener-Hammerstein model, *IEEE Trans. on Autom. Control*, **1995**, *40*, 287–291.

10 Hicks, J., Mohan, A., Ray, W. H., The optimal control of polymerization reactors, *Can. J. Chem. Eng.*, **1969**, *47*, 590–597.

11 Louie, B. R., Soong, D. S., Optimization of batch polymerization process – Narrowing the MWD. I. Model Simulation, *J. Appl. Polym. Sci.*, **1985**, *30*, 3707–3749.

12 Chang, I.-S., Lai, J.-L., Computation of optimal temperature policy for molecular weight control in a batch polymerization reactor, *Ind. Eng. Chem. Res.*, **1992**, *31*, 861–868.

13 Grubecki, I., Woicik, M., Comparison between isothermal and optimal temperature policy for batch reactor, *Chem. Eng. Sci.*, **2000**, *55*, 5161–5163.

14 Platzer, B., Platzer, B. A., Weissenborn, K.-D., Modellierung der Korngrosse bei der Suspensionspolymerisation von Vinylchlorid in Abhangigkeit von Rezeptur, Betriebsbedingungen und Umsatz, *Plaste Kautsch.*, **1987**, *34*, 399–405.

15 Platzer B., Weissenborn, K.-D., Zur Herausbildung ein – und mehrzelliger PVC-S-Korner, *Plaste Kautsch.*, **1987**, *34*, 147–152.

16 Mihail, R., *Modele cinetice de polireactii*, E. S. E., Bucuresti, **1986** q.

17 Henschel, G., Platzer, B., Weickert, G., Prozesskinetik und Sorption – Ein Modell zur Beschreibung der Suspensionspolymerisation von Vinylchlorid bis zum Grenzumsatz, *Plaste Kautsch.*, **1985**, *32*, 329–333.

18 Ugelstadt J., Mark, P. C., Hansen, F., Kinetics and mechanism of vinyl chloride polymerization, *Pure & Appl. Chem.*, **1981**, *53*, 323–363.

19 Coleman, Th., Branch, M. A., Grace, A., Optimization Toolbox, ver. 2.0, The MathWorks Inc., **1999**.

20 Goldberg, D. E., *Genetic Algorithms in Search, Optimization and Machine Learning*, Addison-Wesley, **1989**.

21 Holland, J. H., *Adaptation in Natural and Artificial Systems*. Ann Arbor: MIT Press, **1975**.

22 Gerdes, I. S., Application of genetic algorithms to the problem of free routing for aircraft. In: *Proceedings of the First IEEE Conference on Evolutionary Computation*, Volume 2, Orlando, Florida, 536–541, **1994**.

23 Petry, F. E., Buckles, B. P., Prabhu, D., George, R., Srikanth, R., Fuzzy clustering with genetic search. In: *Proceedings of the First IEEE Conference on Evolutionary Computation*, Volume I, Orlando, Florida, 46–50, **1994**.

24 Davidor, Y., *Genetic Algorithms and Robotics: a heuristic strategy for optimization*. Singapore: World Scientific, **1991**.

25 Karr, C. L., An adaptive system for process control using genetic algorithms. In: *IFAC Artificial Intelligence in Real-time Control*, Delft, The Netherlands, 329–334, **1992**.

26 Kacprzyk, J., A modified genetic algorithm for multistage control of a fuzzy system. In: *Proceedings EUFIT'95*, Volume 1, Aachen, Germany, 463–466, **1995**.

27 Linkens, D. A., Nyongesa, H. O., Genetic algorithms for fuzzy control. Part I: Offline system development and application, *IEE Proceedings on Control Theory Applications*, **1995**, *142*(3), 161–175.

28 Linkens, D. A., Nyongesa, H. O., Genetic algorithms for fuzzy control-part 2: Online system development and application, *IEE Proceedings on Control Theory Applications*, **1995**, *142*(3), 177–185.

29 Surmann, H., Kanstein A., Goser, K., Self-organizing and genetic algorithms for an automatic design of fuzzy control and decision systems. In: *Proceedings of First European Congress on Intelligent*

Techniques and Soft Computing, Volume 2, 1097–1104, **1993**.

30 Nagy, Z., Agachi, S., Productivity optimization of the PVC batch suspension reactor using genetic algorithm. In: *Proceedings, 2nd Conference on Process Integration, Modeling and Optimization for Energy Saving and Pollution Reduction, PRES«99*, Budapest, Hungary, 487–492, **1999**.

31 Poloski, A. P., Kantor, J. C., Application of model predictive control to batch processes, *Computers and Chemical Engineering*, **2003**, *27*, 913–926.

32 Dulce, C., Silva, C. M., Oliveira, M. C., Optimization and nonlinear model predictive control of batch polymerization systems, *Computers and Chemical Engineering*, **2002**, *26*, 649–658.

33 Nagy, Z. K., Mahn, B., Franke, R., Allgower, F., Nonlinear model predictive control of batch processes: an industrial case study. In: *Proceedings, 16th IFAC World Congress*, Prague, July 4–8, on CD, **2005**.

34 Nagy, Z., Agachi, S., Model predictive control of a PVC batch reactor, *Computers and Chemical Engineering*, **1997**, *21*(6), 571–591.

35 Valappil, J., Georgakis, C., Accounting for batch reactor uncertainty in the nonlinear MPC of end-use properties, *AIChE Journal*, **2003**, *49*(5), 1178–1192.

36 Nagy, Z., Braatz, R., Robust nonlinear model predictive control of batch processes, *AIChE Journal*, **2003**, *49*(7), 1776–1786.

37 Özcan, G., Hapoglu, H., Alpbaz, M., Generalized predictive control of optimal temperature profiles in a polystyrene polymerization reactor, *Chemical Engineering and Processing*, **1998**, 37, 125–130.

38 Özcan, G., Kothare, M., Georgakis, C., Control of a solution copolymerization reactor using multi-model predictive control, *Chemical Engineering Science*, **2003**, 58, 1207–1221.

39 Alhamad, B., Romagnoli, J., Gomes, V., On-line multi-variable predictive control of mass and particle size distribution in free-radical emulsion copolymerization, *Chemical Engineering Science*, **2005**, 60, 6596–6606.

40 Shi, D., El-Farra, N., Li, M., Mhaskar, P., Christofides, P., Predictive control of particle size distribution in particulate processes, *Chemical Engineering Science*, **2006**, 61, 268–281.

41 Garcia, C. E., Quadratic dynamic matrix control of nonlinear processes. An application to a batch reaction process, AIChE Annual Meeting, San Francisco, California, **1984**.

42 Gattu, G., Zafiriou, E., nonlinear quadratic dynamic matrix control with state estimation, *Ind. End. Chem. Res.*, **1992**, *31*, 1096–1104.

43 Nagy, Z., Agachi, S., Reglarea predictivă după model, neliniară a procesului discontinuu de fabricare a PVC, *Zilele Academice Timişene*, Timişoara, vol. 2, 39–42, **1995**.

44 Kalra, L., Georgakis, C., Effect of process nonlinearity on the performance of linear model predictive controllers for the environmentally safe operation of a fluid catalytic cracking unit, *Ind. Eng. Chem. Res.*, **1994**, *33*, 3063–3069.

45 Bequette, B. W., Nonlinear control of chemical processes – a review, *Ind. Eng. Chem. Res.*, **1991**, *30*, 1391–1413.

46 Roels, J. A., Mathematical models and the design of biochemical reactors, *J. Chem. Tech. Biotechnol.*, **1982**, *32*, 59–72.

47 Volesky, B., Yerushalmi, L., Luong, J. H. T., Metabolic-heat relation for aerobic yeast respiration and fermentation, *J. Chem. Tech. Biotechnol.*, **1982**, *32*, 650–659.

48 Cho, G. H., Choi, C. Y., Choi, Y. D., Han, M. H. Ethanol production by immobilised yeast and its CO_2 gas effects in a packed bed reactor, *J. Chem. Tech. Biotechnol.*, **1982**, *32*, 959–967.

49 Campello, R. J., Von Zuben, F. J., Amaral, W. C., Meleiro, C. A., Maciel Filho, R., Hierarchical fuzzy models within the framework of orthonormal basis functions and their applications to bioprocess control, *Chemical Engineering Science*, **2003**, 58, 4259–4270.

50 Aiba, S., Shoda, M., Nagatani, M., Kinetics of product inhibition in alcoholic fermentation, *Biotechnol. Bioeng.*, **1968**, *10*, 846–864.

51 Sevella, B., *Bioengeneering operations*, Technical University of Budapest, Ed. Tankonykiado, Budapest, **1992**.

52 Godia, F., Casas, C., Sola, C., Batch alcoholic fermentation modelling by simultaneous integration of growth and fermentation equations, *J. Chem. Tech. Biotechnol.*, **1988**, *41*, 155–165.

53 Nagy, Z., Agachi, S., Dynamic modelling and model predictive control of a fermentation bioreactor. In: *Proceedings on the International Chemistry Show of the Romanian Society of Chemical Engineers, SiChem«98*, October 20–23, Bucuresti, 350–359, **1998**.

54 Wang, F.-S., Sheu, J.-W., Multiobjective parameter estimation problems of fermentation processes using a high ethanol tolerance yeast, *Chem. Eng. Sci.*, **2000**, *55*, 3685–3695.

55 Zhu, G.-Y., Zamamiri, A., Henson, M. A., Hjorts, M. A., Model predictive control of continuous yeast bioreactors using cell population balance models, *Chem. Eng. Sci.*, **2000**, *55*, 6155–6167.

56 Mjalli, F. S., Al-Asheh, S., Neural-networks-based feedback linearization versus model predictive control of continuous alcoholic fermentation process, *Chemical Engineering and Technology*, **2005**, *28*(10), 1191–1200.

57 Ramaswamy, S. Cutright, T. J., Qammar, H. K., Control of a continuous bioreactor using model predictive control, *Process Biochemistry*, **2005**, *40*(8), 2763–2770.

58 Pollard, J. F., Broussard, M. R., Garrison, D. B., San, K. Y., Process identification using neural networks, *Computers and Chemical Engineering*, **1992**, *16*(4), 253–270.

59 Chen, J., Bruns, D. P., WaveARX neural network development for system identification using a systematic design synthesis, *Ind. Eng. Chem. Res.*, **1995**, *34*(12), 4420–4435.

60 Kramer, M. A., Autoassociative neural networks, *Computers and Chemical Engineering*, **1992**, *16*(4), 313–328.

61 Mavrovouniotis, M. L., Chang, S., Hierarchical neural networks, *Computers and Chemical Engineering*, **1992**, *16*(4), 347–369.

62 Bhat, N. V., McAvoy, T. J., Determining model structure for neural models by network stripping, *Computers and Chemical Engineering*, **1992**, *16*(4), 271–281.

63 Hansen, L. K., Pedersen, M. W., Controlled growth of cascade correlation nets. In: Marino, M., Morasso, P. G. (Eds), *Proceedings, ICANN'94*, Sorrento, Italy, 797–800, **1994**.

64 Nagy, Z., Agachi, S., Using optimal brain surgeon for the determination of the topology of artificial neural networks used in the dynamic modeling of chemical processes, *The XIth Romanian International Conference on Chemistry and Chemical Engineering, RICCE-11*, Bucharest, full paper on CD, **1999**.

65 Nagy, Z., Agachi, S., Nonlinear model predictive control of a continuous fermentation reactor using artificial neural networks, *Automatic Control and Testing Conference, Cluj-Napoca*, 23rd–24th May, Section A2, 235–240, **1996**.

66 Nagy, Z., Agachi, S., Nonlinear model predictive control of a continuous fermentation reactor using neural networks. In: *Proceedings of the Scientific Conference of Artificial Intelligence in Industry, AIII«98*, Slovakia, 207–215, **1998**.

67 Qin, S. J., Badgwell, T. A., An overview of nonlinear model predictive control applications. In: Allgower, F., Zheng, A. (Eds), *Nonlinear Predictive Control*, Birkhauser, Basel, pp. 369–392, **2000**.

68 Richalet J., Industrial applications of model based predictive control, *Automatica*, **1993**, *29*(5), 1251–1274.

69 Biegler, L. T., Rawlings, J. B., Optimisation approaches to nonlinear model predictive control, *Proceedings of Conference on Chemical Process Control*, South Padre Island, Texas, 543–571, **1991**.

70 Chen, H., Allgower, F., Nonlinear model predictive control schemes with guaranteed stability. In: Berber, C., Kravaris, R. (Eds), *Nonlinear Model Based Process Control*, Kluwer Academic Publishers, Dordrecht, pp. 465–494, **1998**.

71 Henson, A. M., Nonlinear model predictive control: current status and future directions, *Computers and Chemical Engineering*, **1998**, *23*, 187–201.

72 Patwardhan, A. A., Edgar, T. F., Nonlinear model predictive control of a packed distillation column, *Ind. Eng. Chem. Res.*, **1993**, *32*, 2345–2356.

73 Sistu, P. B., Bequette, B. W., Nonlinear model-predictive control: closed-loop stability analysis, *AIChE Journal*, **1996**, *42*, 3388–3202.

74 Findeisen, R., Allgower, F., Diehl, M., Bock, H. G., Schloder, J. P., Nagy, Z., Computational feasibility of nonlinear model predictive control, *European Control Conference*, Porto, Portugal, **2001**.

75 Khowinij, S., Bian, S., Henson, M. A., Belanger, P., Megan, L., Reduced order modeling of high purity distillation columns for nonlinear model predictive control, in *Proceedings of the 2004 American Control Conference*, Vol. 5, 4237–42, **2004**.

76 Bian, S., Khowinij, S., Henson, M. A., Belanger, P., Megan, L., Compartmental modeling of high purity air separation columns, *Computers and Chemical Engineering*, **2005**, *29*(10), 2096–2109.

77 Nagy, Z., Agachi, S., Real-time implementation of nonlinear model predictive control to a pilot distillation column, *Control Engineering and Applied Informatics*, **2001**, *3*(4), 5–18.

78 Waller, J. B., Boling, J. M., Multi-variable nonlinear MPC of an ill-conditioned distillation column, *Journal of Process Control*, **2005**, *15*(1), 23–29.

79 Chen, H., Allgower, F., A quasi-infinite horizon nonlinear model predictive control scheme with guaranteed stability, *Automatica*, **1998**, *34*(10), 1205–1218.

80 Skogestad. S., Morari, M., Understanding the dynamic behavior of distillation columns, *Ind. Eng. Chem. Res.*, **1988**, *27*, 1848–1862.

81 Skogestad, S., Dynamics and control of distillation columns – A tutorial introduction, *Trans. IchemE*, **1997**, *75*(A).

82 Skogestad, S., Dynamics and control of distillation columns – A critical survey, *Modeling, Identification and Control*, **1997**, *18*, 177–217.

83 Liu, C. J., Yuan, X. G., Yu, K. T., Zhu, X. J., A fluid-dynamic model for flow pattern on a distillation tray, *Chem. Eng. Sci.*, **2000**, *55*, 2287–2294.

84 Rehm, A., Allgower, F., Nonlinear H_{inf}-control of a high purity distillation column, *UKACC International Conference on Control'96*, 1178–1183, **1996**.

85 Kienle, A., Low-order dynamic models for ideal multicomponent distillation processes using nonlinear wave propagation theory, *Chem. Eng. Sci.*, **2000**, *55*, 1817–1828.

86 Findeisen, R., Allgower, F., Nonlinear model predictive control for index-one DAE systems. In: Allgower, F., Zheng, A. (Eds), *Nonlinear Predictive Control*, Birkhauser, **2000**.

87 Findeisen R., Allgower, F., A nonlinear model predictive control scheme for the stabilization of setpoint families, *Benelux Q. J. Automatic Control*, **2000**, *41*(1), 37–45.

88 Diehl, M., Bock, H. G., Schlöder, J. P., Findeisen, R., Nagy, Z., Allgöwer, F., Real-time optimization and nonlinear model predictive control of processes governed by differential-algebraic equations, *Journal of Process Control*, **2002**, *12*, 577–585.

89 Bock H. G., Eich, E., Schloder, J. P., Numerical solution of constrained least squares boundary value problems in differential-algebraic equations. In: Strehmel, K. (Ed.), *Numerical Treatment of Differential Equations*, Teubner, **1988**.

90 Leineweber D. B., *Efficient reduced SQP methods for the optimization of chemical processes described by large sparse DAE models*, Ph.D. thesis, University of Heidelberg, **1998**.

91 Schulz, V. H., Bock, H. G., Steinbach, M. C., Exploiting invariants in the numerical solution of multipoint boundary value problems for DAEs, *SIAM J. Sci. Comp.*, **1998**, *19*, 440–467.

92 Bock, H. G., Bauer, I., Leineweber, D., Schloder, J., Direct multiple shooting methods for control and optimization in engineering. In: Keil, F., Mackens, W., Vos, H., Werther, J. (Eds), *Scientific Computing in Chemical Engineering II*, Volume 2, Springer, **1999**.

93 Bock, H. G., Diehl, M., Leineweber, D., Schloder, J., A direct multiple shooting method for real-time optimization of nonlinear DAE processes. In: Allgower, F., Zheng, A. (Eds), *Nonlinear Predictive Control*, Birkhauser, **2000**.

94 Bock, H. G., Plitt, K. J., A multiple shooting algorithm for direct solution of optimal control problems. In: *Proceedings, 9th IFAC World Congress*, Budapest, Hungary, Pergamon Press, **1984**.

95 Leineweber, D.B, Efficient reduced SQP methods for the optimization of chemical processes described by large sparse DAE models, Vol. 613 of Fortschr.-Ber. VDI Reihe 3, Verfahrenstechnik. VDI Verlag, Düsseldorf, **1999**.

96 Bock, H. G., Diehl, M., Schloder, J. P., Allgower, F., Findeisen, R., Nagy, Z., Real-time optimization and nonlinear model predictive control of processes governed by differential-algebraic equations. In: *Proceedings of the International Symposium on Advanced Control of Chemical Processes, ADCHEM 2000*, Pisa, Italy, 695–703, **2000**.

97 Nagy, Z., Findeisen, R., Diehl, M., Allgower, F., Bock, H. G., Agachi, S., Schloder, J. P., Leineweber, D., Real-time feasibility of nonlinear predictive control for large scale processes – a case study. In: *Proceedings of the American Control Conference*, Chicago, Illinois, 4249–4254, **2000**.

98 Allgower F., Badgwell, T. A., Quin, J. S., Rawlings, J. B., Wright, S. J., Nonlinear predictive control and moving horizon estimation – An introductory overview. In: Frank, P. M. (Ed.), *Advances in Control, Highlights of ECC'99*, 391–449, Springer, **1999**.

99 Lee, J. H., Ricker, N. L., Extended Kalman filter based nonlinear model predictive control, *Ind. Eng. Chem. Res.*, **1994**, 33, 1530–1541.

100 Michalska, H., Mayne, D. Q., Moving horizon observers and observer-based control. *IEEE Trans. Automat. Contr.*, **1995**, 40(6), 995–1006.

101 Oisiovici, R. M., Cruz, S. L., State estimation of batch distillation columns using an extended Kalman filter, *Chem. Eng. Sci.*, **2000**, 55, 4667–4680.

102 Casavola, A., Mosca, E., Predictive reference governor with computational delay, *European Journal of Control*, **1998**, 4(3), 241–248.

103 Von Wissel, D., Nikoukhah, R., Campbell, S. L., Delebecque, F., Effects of computational delay in descriptor-based trajectory tracking control, *International Journal of Control*, **1997**, 67, 251–273.

104 Nagy, Z., Agachi, S., Allgower, F., Findeisen, R., Nonlinear model predictive control of a high purity distillation column, *14th International Congress of Chemical and Process Engineering CHISA«2000*, 27–31 August, Praha, full text on CD, Paper No. 0998 (P3.12), **2000**.

105 Nagy, Z., Agachi, S., Allgower, F., Findeisen, R., Diehl, M., Bock, H. G., Schloder, J. P., Using genetic algorithm in robust nonlinear model predictive control, *European Symposium on Computer Aided Process Engineering-11, ESCAPE-11*, Denmark, 27–30 May, **2001**.

106 MacKay, D. J. C., Bayesian interpolation, *Neural Computation*, **1992**, 4(3), 415–447

107 Fileti, A. F., Cruz, S. L., Pereira, J. A. F.R., Control strategies analysis for a batch distillation column with experimental testing, *Chemical Engineering and Processing*, **2000**, 39, 121–128.

108 Hapoglu, H., Karacan, S., Koca, Z. S. E., Alpbaz, M., Parametric and nonparametric model based control of a packed distillation column, *Chemical Engineering and Processing*, **2001**, 40, 537–544.

109 Karacan, S., Application of a non-linear long range predictive control to a packed distillation column, *Chemical Engineering and Processing*, **2003**, 42, 943–953.

110 Alpbaz, A., Karacan, S., Cabbar, Y., Hapoglu, H., Application of model predictive control and dynamic analysis to a pilot distillation column and experimental verification, *Chemical Engineering Journal*, **2002**, 88(1–3), 163–174.

111 Diehl, M., Findeisen, R., Schwarzkopf, S., Uslu, I., Allgower, F., Bock, H. G., Gilles, E. D., Schloder, J. P., An efficient algorithm for nonlinear model predictive control of large-scale systems. II.

Experimental evaluation for a distillation column, *Automatisierungstechnik*, **2003**, *51*(1), 22–29.

112 Avidan, A. A., Shinnar, R., Development of catalytic cracking technology. a lesson in chemical reactor design, *Ind. Eng. Chem. Res.*, **1990**, *29*, 931–942.

113 Advanced Process Control Handbook VII, *Hydrocarbon Processing*, **1992**, May, pp. 122–126.

114 Arbel, A., Huang. Z., Rinard, I., Shinnar, R., Dynamics and control of fluidized catalytic crackers. 1. Modeling of the current generation of FCCUs, *Ind. Eng. Chem. Res.*, **1995**, *34*, 1228–1243.

115 Arbel, A., Huang. Z., Rinard, I., Shinnar, R., Partial control of FCC units: Input multiplicities and control structures, *AIChE Annual Meeting*, St. Louis, MO, **1993**.

116 Cristea, M. V., Toma, L., Agachi, S. P., Neural networks used for model predictive control of the fluid catalytic cracking unit, *7th World Congress of Chemical Engineering*, Glasgow, June, **2005**.

117 Vieira, W. G., Santos, V. M. L., Carvalho, F. R., Pereira, J. A. F.R., Fileti, A. M. F., Identification and predictive control of a FCC unit using a MIMO neural model, *Chemical Engineering and Processing*, **2005**, *44*, 855–868.

118 McFarlane, R. C., Reineman, R. C., Bartee, J. F., Georgakis, C., Dynamic simulator for a Model IV fluid catalytic cracking unit, *Computers and Chemical Engineering*, **1993**, *17*, 275–300.

119 Cristea, M. V., Marinoiu, V., Agachi, S. P., Aspects of traditional and advanced control of the fluid catalytic cracking unit, *Revue Roumaine de Chimie*, **2002**, *47*(10–11), 1127–1132.

120 Emad, E. A., Elnashaie, S. E. H., Nonlinear model predictive control of industrial type IV fluid catalytic cracking units for maximum gasoline yield, *Ind. Chem. Eng. Res.*, **1997**, *36*, 389–398.

121 Karla, L., Georgakis, C., Effect of process nonlinearity on performance of linear model predictive controllers for environmentally safe operation of a fluid catalytic cracking unit, *Ind. Eng. Chem. Res.*, **1994**, *33*, 3063–3069.

122 Garcia, C. E., Prett, M. P., Morari, M., Model predictive control: theory and practice – a survey, *Automatica*, **1989**, *25*, 335–348.

123 Chen, H., Allgower, F., Nonlinear model predictive control schemes with guaranteed stability, *NATO Advanced Study Institute on Nonlinear Model Based Process Control*, August 10–20, Antalya, Turkey, 1–28, **1997**.

124 Cristea, M. V., Agachi, S. P., Simulation and model predictive control of the soda ash rotary calciner, *Control Engineering and Applied Informatics*, **2001**, *3*(4), 19–26.

125 Cristea, M. V., Agachi, S. P., Marinoiu, M. V., Simulation and model predictive control of a UOP fluid catalytic cracking unit, *Chemical Engineering and Processing*, **2003**, *42*, 67–91.

126 Weekman, V. W., Jr., Nace, D. M., Kinetics of catalytic cracking selectivity in fixed, moving, and fluid bed reactors, *AIChE Journal*, **1970**, *16*, 397–404.

127 Cristea, M. V., Agachi, S. P., Dynamic simulator for a UOP model fluid catalytic cracking unit, *Studia Universitatis "Babeş-Bolyai" Ser. Chemia*, **1997**, *42*, 97–102.

128 Cristea, M. V., Marinoiu, V., Agachi, S. P., *Reglarea predictivă după model a instalaţiei de cracare catalitică*, ISBN: 973–686–412-X, Editura Casa Cărţii de Ştiinţă, **2003**, Cluj-Napoca, Romania.

129 Hovd, M., Skogestad, S., Controllability analysis for the fluid catalytic cracking process, *AIChE Annual Meeting*, November 17–22, Los Angeles, **1991**.

130 Lee, E., Groves, F. R., Jr., Mathematical model of the fluidized bed catalytic cracking plant, *Transaction of The Society for Computer Simulation*, **1985**, *3*, 219–236.

131 Shah, Y. T., Huling, G. P., Paraskos, J. A., McKinney, J. D., A kinematic model for an adiabatic transfer line catalytic cracking reactor, *Ind. Eng. Chem., Process Des. Dev.*, **1977**, *16*, 89–94.

132 Kurihara, H., *Optimal control of fluid catalytic cracking process*, Ph.D. Thesis, **1967**, MIT.

133 Morari, M., Huq, I., Agachi, S. P., Bomberger, J., Donno, B., Zengh, A.,

Modeling and control studies on Model IV FCCUs, *Joint Research between Chevron Research and Technology Company and California Institute of Technology, 1993.*

134 Rhemann, H., Schwartz, G., Badgwell, T., Darby, M., White, D., On-line FCCU advanced control and optimization, *Hydrocarbon Processing,* **1989**, June 6, pp. 64–69.

135 McDonald, G., Harkins, B., Maximizing Profits by Process Optimization, *1987 NPRA Annual Meeting,* March 29–31, San Antonio, Texas, **1987**.

136 Yang, S., Wang, X., McGreavy, C., A multivariable coordinated control system based on predictive control strategy for FCC reactor-regenerator system, *Chemical Engineering Science,* **1996**, *51,* 2977–2982.

137 Kookos, I. K., Perkins, J. D., Regulatory control structure selection of linear systems, *Computers and Chemical Engineering,* **2002**, *26,* 875–887.

138 Cristea, M. V., Agachi, S. P., Controllability analysis of a Model IV FCCU, *12th International Congress of Chemical and Process Engineering CHISA,* **1996**, Prague, 10.

139 Semino, D., Scali, C., A method for robust tuning of linear quadratic optimal controllers, *Ind. Eng. Chem. Res.,* **1994**, *33,* 889–895.

140 Lee, J. H., Yu, Z. H., Tuning of model predictive controllers for robust performance, *Computers Chem. Engng.,* **1994**, *18,* 15–37.

141 Patwardhan, A. A., Rawlings, J. B., Edgar, T. F., Nonlinear MODEL PREDICTIVE CONTROL, *Chem. Eng. Comm.,* **1990**, *87,* 123–141.

142 Jia, C., Rohani, S., Jutan, A., FCC unit modeling, identification and model predictive control, a simulation study, *Chemical Engineering and Processing,* **2003**, *42,* 311–325.

143 Richalet, J., Industrial applications of model based predictive control, *Automatica,* **1993**, *29,* 1251–1274.

144 Young, J. C. C., Baker, R., Swartz, C. L. E., Input saturation effects in optimizing control-inclusion within a si-

multaneous optimization framework, *Computers and Chemical Engineering,* **2004**, *28,* 1347–1360.

145 Augier, F., Coumans, W. J., Hugget, A., Kaasschieter, E. F., On the risk of cracking in clay drying, *Chemical Engineering Journal,* **2002**, *86,* 133–138

146 Su, S.-L., Modeling of multiphase moisture transfer and induced stress in drying clay bricks, *Applied Clay Science,* **1997**, *12,* 189–207.

147 Qin, J. S., Badgwell, T. A., A survey of industrial model, predictive control technology. *Control Engineering Practice,* **2003**, *11*(7), 733–764.

148 Cristea, M. V., Baldea M., Agachi, S. P., Model predictive control of an industrial dryer, *ESCAPE-10,* Florence, **2000**, 271–276.

149 Cristea, M. V., Roman, R., Agachi, S. P., Neural networks based model predictive control of the drying process, *European Symposium on Computer Aided Process Engineering-13,* Lappeenranta, Finland, 1–4 June, pp. 389–394, **2003**.

150 Khalfi, A., Blanchart, P., Desorption of water during the drying of clay minerals. Enthalpy and entropy variation, *Ceramics International,* **1999**, *25,* 409–414.

151 Dufour, P., Toure, Y., Blanc, D., Laurent, P., On nonlinear distributed parameter model predictive control strategy: online calculation time reduction and application to an experimental drying process, *Computers and Chemical Engineering,* **2003**, *27,* 1533–1542.

152 van Meel, D. A., Adiabatic convection batch drying with recirculation of air, *Chem. Engng. Sci.,* **1958**, *9,* 36–44.

153 Krischer, O., Kast, W., *Die wissenschaftlichen Grundlagen der Trocknungstechnik,* Springer-Verlag, **1992**.

154 Cristea, M., Agachi, S., Göbel, C., Sensitivity analysis of an industrial dryer, *13th International Congress of Chemical and Process Engineering CHISA,* 23–28 August, Praha, **1998**, on CD-ROM.

155 Perry, R. H., Green, D. W., *Perry's Chemical Engineers' Handbook,* 7th edition, McGraw-Hill, **1999**.

156 Garcia, C. E., Prett, M. P., Morari, M., Model predictive control: theory and practice – a survey, *Automatica*, **1989**, 25(3), 335–348.

157 Cristea, M. V., Agachi, S. P., Reglarea evoluata a reactorului de carbonatare din instalatia de producerea sodei amoniacale, *Revista Românâ de Informaticâ şi Automaticâ*, **1997**, 7(4), 45.

158 Cristea, M. V., Marinoiu, V., Agachi, S. P., Multivariable model based predictive control of a UOP fluid catalytic cracking unit, *2nd Conference on Process Integration, Modeling and Optimization for Energy Saving and Pollution Reduction*, Budapest, pp. 223–228, **1999**.

159 Cristea, M. V., Toma, L., Agachi, S. P., Neural networks used for model predictive control of the fluid catalytic cracking unit, *7th World Congress of Chemical Engineering*, Glasgow, June **2005**, on CD-ROM.

160 Hagan, M. T., Menhaj, M. H., Training feedforward networks with the Marquardt algorithm, *IEEE Transaction on Neural Networks*, **1994**, 5, 989–1003.

161 Stringer, R., Johnston, P., *Chlorine in the Environment: An Overview of the Chlorine Industry*, Kluwer Academic Publisher, Dordrecht, **2001**.

162 Srinivasan, V., Arora, P., Ramadass, P., Report on the electrolytic industries for the year 2004, *J. Electrochem. Soc.*, **2006**, 153(4), K1–K14.

163 Szep, A., Bandrabur, F., Manea, I., *The electrolysis of natrium chloride solutions by ion exchange membrane process* [in Romanian], Editura CERMI, Iaşi, Romania, **1998**.

164 Oniciu, L., Ilea, P., Popescu, I. C., *Technological Electrochemistry* [in Romanian], Casa Cartii de Stiinta, Cluj-Napoca, Romania, **1995**.

165 Agachi, S ., Imre, A., Socol, I., Oniciu, L., Steady state model of the electrochemical process of brine electrolysis in De Nora amalgam reactors, *Hung. Journal of Industrial Chemistry (HJIC)*, Hungary, **1997**, 25(2), 81–90.

166 Imre, A., *Modeling and control of brine electrolysis by mercury cell process and ion exchange membrane cell process*, PhD Thesis, University "Babes-Bolyai" Cluj, Romania, **1999**.

167 Pokhozhaev, S. J., Mathematical model of electrolysis, *Dokl. Akad. Nauk*, **1993**, 332(6), 690–692.

168 Nagy, Z., A mechanistic model for the calculation of material balance for a diaphragm-type chlorine caustic cell, *J. Electrochem. Soc.*, **1977**, 124(1), 91–95.

169 Imre, A., Dynamic model of De Nora amalgam reactor for brine electrolysis, *13th International Conference of Chemical and Process Engineering, CHISA'98*, Prague, **1998**.

170 Leistra, J. A., Sides, P. J., Voltage components at gas evolving electrodes, *J. Electrochem. Soc.*, **1987**, 134(10), 2442–2446.

171 Kuhn, M., Kreysa, G., Modeling of gas-evolving electrolysis cells, *J. Appl. Electrochem.*, **1989**, 19, 720–728.

172 Faita, G., Longhi, P., Mussini, T., Standard potential of the Cl_2/Cl electrode at various temperatures with related thermodynamic functions, *J. Electrochem. Soc.*, **1967**, 114(4), 340–343.

173 Tilak, B. V., Kinetics of chlorine evolution – A comparative study, *J. Electrochem. Soc.*, **1979**, 126(8), 1343–1348.

174 Oniciu, L., Agachi, S., Bugan, J., Imre, A., *Implementarea calculatorului de process la electroliza solutiei de clorura de sodiu*, research report, UBB Cluj-Napoca, Romania, **1989**.

175 Caldwell, D. L., Production of chlorine. In: Bockris, J.O'M, Conway, B. E., Yeager, E., White, R. E. (Eds), *Comprehensive Treatise of Electrochemistry*, vol. 2, Plenum Press, New York, London, **1981**.

176 Vaaler, L. E., Voltage drop and current distribution using metal anodes in chlorine-caustic cell, *J. Appl. Electrochem.*, **1979**, 9, 21–27.

177 De La Rue, R. E., Tobias, C. W., On the conductivity of dispersions, *J. Electrochem. Soc.*, **1959**, 106, 827–833.

178 Bratu, E. A., *Unit operations in chemical engineering* [in Romanian], vol. I, Editura Tehnicâ, Bucureşti, Romania, **1985**.

179 Vogt, H., The incremental ohmic resistance caused by bubbles adhering to an electrode, *J. Appl. Electrochem.*, **1983**, 13, 87–88.

180 Chin, J. M. K.J., Janssen, L. J. J., van Strelen, S. J. D., Verbunt, J. H. G.,

Sluyter, W. M., Bubble parameters and efficiency of gas evolution for a chlorine, a hydrogen and an oxygen evolving wire electrode, *Electrochem. Acta*, **1988**, *33*(6), 769–771.

181 Chandran, R. R., Chin, D. T., Reactor analysis of a chlor-alkali membrane cell, *Electrochim. Acta*, **1986**, *31*, 39–50.

182 Delmas, F., Production of sodium hydroxide and chlorine by membrane electrolysis, *Rev. Electr. Electron.*, **1995**, *3*, 21–24.

183 Leah, R. T., Brandon, N. P., Vesovic, V., G. H. Kelsall, G. H., Numerical modeling of the mass transport and chemistry of a simplified membrane-divided chlor-alkali reactor, *J. Electrochem. Soc.*, **2000**, *147*(11), 4173–4183.

184 Martens, L., Hertwig, K., Preidt, C., Mathematische Modellierung von Jalousie-Vorelektroden in Membranzellen der Chloralkali-Elektrolyse, *Chem.-Ing.-Tech.*, **1992**, *8*(64), 719–722.

185 Bergner, D., *Membrane electrolyzers for the chlor-alkali industry*, Dechema-Monographs, Vol. 123, VCH Verlagsgesellschaft, **1991**.

186 Burney, H. S., Membrane chlor-alkali process. In: *Modern Aspects of Electrochemistry*, Plenum Press, New York, **1993**.

187 Keating, J. T., Behling K. J., Brine, impurities, and membrane chlor-alkali cell performance, *DECHEMA*, **1990**, *123*, 125–139.

188 Ogata, Y., Kojima, T., Uchiyama, S., Yasuda, M., Hine, F., Effects of the brine impurities on the performance of the membrane-type chlor-alkali cell, *J. Electrochem. Soc.*, **1989**, *136*(1), 91–95.

189 Yeager, H. L., Kipling, B., Dotson, R. L., Sodium ion diffusion in Nafion ion exchange membranes. *J. Electrochem. Soc.*, **1980**, *127*, 303–307.

190 Yeager, H. L., O'Dell, B., Twardowski, Z., Transport properties of Nafion membranes in concentrated solution environments, *J. Electrochem. Soc.*, **1982**, *129*(1), 85–89.

191 Oniciu, L., Agachi, S., Socol, J., Imre, A., *Modelul matematic al reactorului de electroliza cu diafragma*, UBB Cluj, research report, Romania, **1990**.

192 Agachi, S., Imre, A., *Control Strategies for Brine Electrolysis by Ion Exchange Membrane Cell Process*, ESCAPE-10, Elsevier Science, **2002**, 289–294.

193 Imre, A., Agachi, S., *Model predictive control of brine electrolysis by ion exchange membrane reactors*, CHISA **2000**, Prague, **2000**, on CD.

Index

Agachi/Nagy/Cristea/Imre-Lucaci. *Model Based Control*
Copyright © 2006 WILEY-VCH Verlag GmbH & Co. KGaA, Weinheim
ISBN: 3-527-31545-4